TELESCOPIC HYDRAULIC GANTRY SYSTEMS

TELESCOPIC HYDRAULIC GANTRY SYSTEMS

David Duerr, P.E.

2DM Associates, Inc.
Houston, Texas

LEVARE PRESS

This book is printed on acid-free paper. ∞

Telescopic Hydraulic Gantry Systems

ISBN: 978-0-615-75016-3

10 9 8 7 6 5 4 3

Contents

6

Engineering Methods and Practices 139

7

Lift Planning and Operations.. 197

About the Author

David Duerr, P.E. is president of 2DM Associates, Inc., his consulting engineering practice in Houston, Texas. Mr. Duerr is a specialist in engineering for heavy lifting and transportation and began working in the field in 1974. He has been working with hydraulic gantry systems since 1977, during which time he has engineered numerous heavy and specialized lifts using telescopic hydraulic gantries and has designed gantry systems of both the bare cylinder and telescopic boom types.

Mr. Duerr holds a Bachelor of Engineering degree from Pratt Institute, a Master of Science in Civil Engineering degree from the University of Houston, and is a licensed professional engineer in many states. He is a member of the American Society of Civil Engineers, the American Society of Mechanical Engineers, the Society of Automotive Engineers and the American Council of Engineering Companies, and is listed in a number of biographical references, including *Who's Who in America* and *Who's Who in Science and Engineering*. Mr. Duerr served on the SC&RA Jacking Systems Task Force, which published in 1996 the first industry guidelines for the design and use of telescopic hydraulic gantries, the SC&RA Telescopic Hydraulic Gantries Task Force, which developed an updated guide in 2004, and currently serves on the ASME B30.1 Subcommittee. In 2012, Mr. Duerr was awarded the ASME Safety Codes and Standards Medal for his work with ASME in the development of lifting equipment design and safety standards.

PREFACE

The telescopic hydraulic gantry system is a comparatively new type of equipment in the speciality lifting industry. Widely available as a commercial product only since the early 1980s, telescopic hydraulic gantries have quickly grown in popularity for lifting work in both industrial maintenance operations and in heavy construction.

The distinguishing characteristic of the telescopic hydraulic gantry system is its variable height. The more common gantry crane consists of a structural frame (the gantry) of fixed height and span. The lifting of a load is accomplished by means of a trolley-mounted hoist that rides along the gantry. The telescopic hydraulic gantry system, on the other hand, is built up of two or more extendible legs that are spanned by one or more header beams to which the load to be lifted is rigged. Lifting and lowering the load is accomplished by extending and retracting the legs. Further, the legs can be assembled with various header and cross beam arrangements, thus allowing the gantry system to be configured with a span that is a custom fit to the job at hand.

Although their use at present is most common in the United States, telescopic hydraulic gantry systems are now used by rigging contractors worldwide. Along with this growth in usage comes the need to disseminate information about gantry systems. Users must have an accurate understanding of the operation and behavior of these systems if they are to be able to plan and execute lifts safely and cost-effectively. Thus, the purpose of this book.

All information previously published about the use of hydraulic gantries is in the form of magazine articles or technical papers. Each publication generally addresses only one or two aspects of gantries. This book brings together into one volume this and much additional previously unpublished material about the design and use of hydraulic gantry systems.

The first chapter presents the history of the development of telescopic hydraulic gantries from specialty products and one-of-a-kind machines built by innovative rigging contractors to the present array of commercial gantry products. Chapter 2 contains a general discussion of hydraulic gantry systems and their uses. This chapter will be of particular value to the reader not already familiar with this equipment. Chapters 3 through 5 present technically intensive discussions of the key elements of telescopic hydraulic gantry design, covering the hydraulic system (Chapter 3), the loads to which a gantry system may be subjected during the performance of a lift (Chapter 4), and analysis of the stability of a gantry system (Chapter 5). Chapters 6 and 7 cover topics needed by the gantry system user, including the design of header beams and track, lift planning and engineering, and lift performance. Chapter 6 is technically focused and is oriented toward the lift planner/engineer, while Chapter 7 provides more practical "nuts and bolts" information.

Example problems are used where appropriate to demonstrate and clarify the principles discussed and to illustrate the use of some of the equations presented. The problems are solved using U.S. customary units (USCU), with international system (SI) equivalent values shown in parentheses where useful for clarity. However, the equations developed throughout this book are dimensionally independent unless otherwise indicated and can be used with either USCU or SI units.

If there is one all-encompassing lesson to be learned here, it is that telescopic hydraulic gantry systems are not cranes and they are not jacks and cribbing. The ways that hydraulic gantries function and the ways they manipulate a supported load are simply different from the means by which cranes or jacks lift and maneuver loads. While some aspects of the planning and execution of a lift with a crane or with jacks and cribbing will translate into lifting with gantries, much of the knowledge needed to plan and safely perform a lift with a gantry system is unique. Recognition of this simple but important fact is the first step toward developing the skills needed to work safely with hydraulic gantries. The need for and value of hands-on experience working with gantry systems is a much repeated theme throughout.

Readers of this book must keep in mind that the use of telescopic hydraulic gantries in the rigging industry is still a relatively young practice. As with the equipment itself, the development of industry standards to regulate the design, manufacture, and use of gantries is also in its infancy. Manufacturers and users of gantry systems have great leeway in what they do and how they do it. Much of the information presented in this book is, therefore, primarily the opinion of the author and is based on the author's experience working with hydraulic gantry systems since the late 1970s. As the use of telescopic gantries grows and more experience is gained within the specialty lifting industry, we must expect that practices in their design and use will continue to evolve.

Many people have contributed information and guidance throughout the writing of this book. Richard Belding, Daniel DeYoung, P.E., Edgar Engler, Michael Greiner, William Hudgins, Roger Johnston, P.E., Carlo Karssen, Gary Lorenz, Max Mayer, Beth O'Quinn, Lance Renshaw, Marvin Wasmund, and John Williams all provided background, reference materials, and photographs for the historical discussion in Chapter 1. All of the gantry manufacturers were very generous with information about their products where needed to describe the uses and functions of gantry systems. However, the most significant contributors have been my co-workers at Williams Crane & Rigging, Inc. back in the 1970s and my many clients from the 1980s to the present who work with hydraulic gantries. This book is the culmination of countless discussions, meetings, planning sessions, and, of course, lifts with telescopic hydraulic gantry systems.

David Duerr, P.E.

Houston, Texas
February 2, 2013

About the Cover

The cover photograph shows the first gantry system with which the author worked on heavy lift projects starting in 1977, a 600-ton capacity system then owned by Williams Crane & Rigging, Inc.

1 The History of Hydraulic Gantries

Telescopic hydraulic gantry systems were born in the early 1960s not as the commercial products available today, but as one-of-a-kind pieces of equipment used by rigging contractors for machinery handling. The basic concepts and designs of these gantry systems came from the minds of imaginative riggers and engineers who needed to solve unusual lifting problems. This first chapter traces the invention and evolution of hydraulic gantry systems from their beginning as specialty one-off pieces of lifting equipment to the diverse array of sophisticated commercial products available today.

1.1 THE FIRST HYDRAULIC GANTRIES

The first pure hydraulic gantry was designed and built in 1963 by Hartley Belding, then the chief engineer of Belding Engineering Company in West Chicago, Illinois. The company was awarded a contract to relocate a manufacturing plant owned by Elkay Sink Company in Chicago, Illinois. Among the equipment items to be moved was a number of large presses. Belding conceived the hydraulic gantry system as a practical method of lifting the tall presses and laying them down to the horizontal position for transport, all while working within the confines of the plant building.

Each leg of this system consisted of three 4" (102 mm) bore single stage hydraulic cylinders. The three cylinders were arranged in a straight line and were welded to a base fabricated from simple structural steel shapes. Although the bases were not fitted with wheels, the system could be moved if necessary by setting the bases on industrial rollers. The system was powered by a gasoline engine pump with an output pressure of 10,000 psi (69,000 kPa). The lifting capacity was 37 1/2 tons (34 tonnes) per leg and the stroke was 8 feet (2.44 meters). Fig. 1.1 is a photograph of the original Belding gantry system at work in 1963 laying down a press.

Later gantries built by Belding used cylinders with larger bore diameters, longer strokes, and capacities of up to 100 tons (91 tonnes) per leg. A variation of the three-cylinder gantry was developed in 1967 in which the cylinders were arranged in a triangle, thus providing a wider footprint and correspondingly greater stability. The first gantry system of this configuration retained the performance specifications of the original gantry: a capacity of 37 1/2 tons (34 tonnes) per leg and a stroke of 8 feet (2.44 meters). Also in 1967, Belding began constructing

Figure 1.1 The First Belding Gantry *(Richard A. Belding)*

gantries with wheels integral to the base, thus allowing the gantry to be more easily moved while supporting a load.

In the years following, Belding designed and constructed a variety of gantry systems of various capacities and lift heights. These gantry systems, all of which were of the single stage, bare cylinder style, were used exclusively by Belding in its rigging contracting business. Today, the company, now a part of Walbridge, owns and operates a variety of multiple stage hydraulic gantry systems with capacities ranging up to 800 tons (725 tonnes).

The first hydraulic gantry system developed and sold as a commercial product was a system manufactured by Modern Hydraulics, Inc. also of West Chicago, Illinois, from the mid 1960s to the early 1970s. This gantry system was of the single stage boom type and was manufactured with capacities ranging from 25 tons (23 tonnes) to 100 tons (91 tonnes) per leg.

The Modern Hydraulics gantry is shown in its simplest form in Fig. 1.2. Each leg consisted of two concentric square structural tubes inside of which was a single stage hydraulic cylinder. The long and narrow base was also fabricated from a structural tube shape, as were the diagonal longitudinal braces. The header beams were simple I-beam or wide-flange shapes.

Longitudinal stability was provided by the long bases and the diagonal braces that supported the bottom section of the boom. The header beam was rigidly bolted to the tops of the two legs, thus creating a rigid frame that provided lateral stability. This rigid frame design presented an operational constraint in that the extension rates of the two legs of the gantry had to be synchronized relatively closely to avoid

Figure 1.2 The Basic Modern Hydraulics Gantry *(Modern Hydraulics, Inc.)*

introducing unwanted bending into the system. The internal hydraulic cylinders acted only in compression to lift the load. The square tube leg sections provided bending stiffness to resist horizontal forces.

The gantry in Fig. 1.2 serves only to lift and lower loads through a vertical stroke of 4 feet (1.22 meters). More sophisticated models were later available that featured bases equipped with wheels for travel with a suspended load and with hydraulically powered lift links that could shift the load from side to side. Fig. 1.3 shows a wheel-mounted Modern Hydraulics gantry system that is outfitted with a side shifting mechanism. Modern Hydraulics discontinued manufacture of these gantry systems in the early 1970s and returned to their core business, the design and manufacture of synchronized jacking systems.

The principal design work of these hydraulic gantries was performed by Robin Renshaw, then president of Modern Hydraulics. Belding Engineering Company, a part owner of Modern Hydraulics at the time, provided input into the concept and design of this product line and was the first buyer of a Modern Hydraulics gantry system.

A variation of the rigid-frame gantry was built by Williams Crane & Rigging, Inc., a lifting and rigging contractor based in Richmond, Virginia. In 1971, Williams Crane was faced with the challenge of unloading from barges the steam supply system components for the North Anna Nuclear Power Plant in Virginia. The remote location and configuration of the barge landing did not allow for the use of any type of land-based lifting equipment. The most practical solution was to ground the barge, lift the components straight up, position the hauling equipment beneath, and then drive the components off the barge.

Figure 1.3 Advanced Modern Hydraulics Gantry *(Modern Hydraulics, Inc.)*

The method chosen to accomplish the lifting utilized a hydraulic gantry system. This system, shown in Fig. 1.4 lifting a steam generator on a later project, consisted of two independent gantries, each of which was built up from two single stage cylinder legs and a pair of wide flange shapes as a header beam. Structural extensions to the cylinders, each 6 feet (1.83 meters) long, could be used as seen in Fig. 1.4 to increase the gantry's overall height. As with the Modern Hydraulics gantry, the header beam was rigidly attached to the cylinders, thus creating a rigid frame that was designed to resist lateral loading and provide lateral stability. One unique design feature of this one-of-a-kind system was the use of pinned connections at the ends of the longitudinal braces and at the bottoms of the cylinder legs. This allowed the gantries to be more easily assembled on site and then dismantled for shipping.

The basic concept of this bare cylinder gantry system was developed and all of the design engineering was performed by Daniel J. DeYoung, P.E., then with the Richmond, Virginia-based consulting engineering firm Torrance, Dreelin, Farthing, and Buford. The system provided a rated load of 600 tons (545 tonnes) and a vertical stroke of 8 feet (2.44 meters). The large bases imposed relatively low bearing pressures to the underlying surface, thus making this system practical for use on light barge decks and on soil.

Williams Crane & Rigging won the Hauling Job of the Year award from the Specialized Carriers & Rigging Association in 1972 for the North Anna project. This was the first of many Job of the Year awards won by contractors over the years for complex or unique projects performed using hydraulic gantry systems.

Figure 1.4 Williams Crane Hydraulic Gantry *(Williams Crane & Rigging, Inc.)*

The next development in the evolution of the hydraulic gantry incorporated two concept changes: the elimination of the dependency of rigid frame behavior to resist lateral loads and the use of multiple stage (telescopic) cylinders. This configuration, like that of the original Belding gantries, used legs with bases that were wide enough relative to their extended heights to provide an adequate stabilizing moment from gravity loads only. The multiple stage cylinders provide a retracted height to extended height ratio that is superior to that possible with single stage cylinders.

The hydraulic legs of the first such gantry were constructed in the late 1970s by Edgar D. Engler, a former employee of Belding Engineering Company who was involved in the development of the Modern Hydraulics gantry. This two-leg gantry system is shown in Fig. 1.5 lifting a 110-ton (100-tonne) generator at a power plant site in St. Louis, Missouri.

Each gantry leg consisted of a base constructed from structural steel shapes on which were mounted three five-stage telescopic hydraulic cylinders of the type normally used to raise the body of a dump truck. A steel header plate joined the tops of the three cylinders and provided a support platform for one or two header beams. This gantry had a retracted height of 8 feet (2.44 meters) and a stroke of 16 feet (4.88 meters) that gave a fully extended height of 24 feet (7.32 meters). The lift capacity was 75 tons (68 tonnes) per leg.

The basic pair of gantry legs was purchased from Engler by Daniel Hamm Heavy Rigging & Transport Company of St. Louis, Missouri, of which Gary V. Lorenz was president. Lorenz then built a power pack for the system and modified the jacks to include safety devices such as counterbalance valves to allow holding

Figure 1.5 The First Telescopic Hydraulic Gantry *(Gary V. Lorenz)*

and controlled lowering of the legs under load. Calculations to determine the capacities of the system were provided by Roger L. Johnston, P.E. through his consulting firm, J&R Engineering Co., Inc. Lorenz and Johnston were previously acquainted at Sargent Engineering and P & H Harnischfeger.

1.2   A PERIOD OF TRANSITION

Lorenz, in partnership with Bill Linden, established Linden Lorenz Rigging Company in Davenport, Iowa, in 1980. This company, later changed to Linden Industrial Inc., performed rigging work using the gantry system rented from Daniel Hamm Heavy Rigging & Transport. Linden Industrial ordered the first gantry system to be built by Riggers Manufacturing Company, a new company formed by Engler, Johnston, and Lorenz to build and sell this new type of multiple stage cylinder gantry. The company's product was based on Engler's original concept, but was built to Linden's order to include wheels, propel cylinders, and a leveling system.

   The Daniel Hamm gantry system (Fig. 1.5) was of a fixed base design. That is, the gantry could not be moved while supporting a load without the use of industrial rollers or some other similar external devices. Riggers Manufacturing improved the product design by introducing a gantry system that featured a wheel-mounted base (Fig. 1.6). This gantry system could be rolled along structural steel track beams while supporting a lifted load. Movement along the track was powered by external propel cylinders (hydraulic cylinders fitted between each gantry leg and

Figure 1.6 Original Riggers Manufacturing Gantry *(Riggers Manufacturing Company)*

the track assemblies). The gantry system was moved by extending or retracting these propel cylinders. (Refer to Section 2.1.3 for a detailed explanation of gantry propel mechanisms.)

Linden Industrial won the SC&RA Rigging Job of the Year award in 1982 for a press installation project performed at the John Deere factory in Davenport, Iowa, using their 500-ton (454-tonne) Riggers Manufacturing gantry system. This was the first Rigging Job of the Year award given for a project in which a hydraulic gantry system was the primary piece of lifting equipment.

A configuration change was made to the Riggers gantry in 1984. Each gantry leg was made with four cylinders, rather than three (Fig. 1.7). By 1984, three different models were produced with capacities ranging from 100 tons (91 tonnes) to 300 tons (272 tonnes) per leg.

The commercial gantry business became a two-company industry in 1983. Gary Lorenz left Riggers Manufacturing Company and founded 4-Point Lift Systems, Inc. (often referred to simply as Lift Systems) in Davenport, Iowa. Lift Systems designed and began manufacturing a line of bare cylinder gantries using only one or two cylinders per leg. These models were most commonly used in arrangements of four legs (as implied by the company name). Depending on the requirements of the lift to be made, the Lift System could be set up as two independent gantries (four legs, two header beams), as is seen in Fig. 1.8, or as one four-leg system with an arrangement of header beams and cross beams that tied all four legs together (Fig. 1.9). This configuration flexibility served the company as a major marketing point.

Roger Johnston left Riggers Manufacturing Company in 1984 and, with Ron McCarthy, founded Hydratech Systems, Inc. in Milwaukee, Wisconsin, to

Figure 1.7 Riggers Manufacturing Company Four-Cylinder Gantry Legs
(Riggers Manufacturing Company)

manufacture hydraulic gantries. Johnston was also associated at the time with Hydra-Power Products, Inc., a manufacturer of hydraulic cylinders.

Hydratech produced two styles of gantry systems. The company's basic product was a bare cylinder model (Fig. 1.10) that used one multiple stage cylinder per leg. Each gantry leg was fitted with a hydrostatic planetary drive (integral drive) propel system that provided continuous travel along a set of steel track beams. The company also produced a telescopic boom gantry system (Fig. 1.11). Each gantry leg of this model consisted of a long structural base on which were supported two telescopic booms. Extension and retraction of the booms was controlled by two

Figure 1.8 500-ton (454-tonne) Capacity 4-Point Lift System *(Lift Systems, Inc.)*

Figure 1.9 4-Point Lift System with Four-Beam Arrangement *(Lift Systems, Inc.)*

cylinders located outboard of the booms. The booms were equipped with locking devices that could support the load mechanically in the event of a hydraulic failure. Hydratech discontinued the manufacture of gantry systems in 1986.

In the mid-1980s, Johnston's Hydra-Power company produced a small number of bare cylinder hydraulic gantry systems, as well as providing hydraulic cylinders to other companies, including Riggers Manufacturing. Hydra-Power also discontinued operations around 1986.

With this discontinuance of gantry manufacturing at Hydra-Power Products, Johnston returned to J&R Engineering. Johnston had founded J&R in 1978 as a consulting engineering firm serving the crane and rigging industry. In 1986, Johnston changed the company's direction and J&R Engineering became the newest manufacturer of telescopic hydraulic gantries.

Figure 1.10 Hydratech Bare Cylinder Gantry *(Roger L. Johnston, P.E.)*

Figure 1.11 Hydratech Telescopic Boom Gantry *(Roger L. Johnston, P.E.)*

1.3   THE CURRENT GANTRY INDUSTRY

J&R Engineering resurrected the telescopic boom gantry of the Modern Hydraulics style. Like the Modern Hydraulics gantries, the J&R products used square structural tubing as the boom sections and placed the hydraulic cylinder inside the boom. Like the Hydratech gantry, the J&R gantry leg design also provided mechanical boom locks. However, where the Hydratech design used stepped latches that could lock the boom sections together only at set intervals along the length of extension, the J&R gantry used cam locking mechanisms between boom sections that enabled the load to be supported mechanically at any position of extension.

As of the late 1980s, three companies remained as the primary manufacturers of commercial gantry products. These are J&R Engineering Company, Inc., Riggers Manufacturing Company, and 4-Point Lift Systems, Inc. They remain today as the only commercial manufacturers of telescopic hydraulic gantry systems based in the United States.

By 1988, J&R Engineering was producing three-stage telescopic boom gantries with cam locks. Early models utilized a three-stage internal cylinder, thus providing a fully powered boom. Later versions of the three-stage boom used a two-stage cylinder and a manual boom section. A 450-ton (408-tonne) J&R boom gantry system is shown in Fig. 1.12.

In 1992, 4-Point Lift Systems added a line of telescopic boom gantry systems to its product mix. The first of these gantries utilized a combination of internal and external cylinders to provide powered extension of all stages of the gantry leg. The lower two or three stages were controlled by two or four external cylinders mounted on the base and tied to the boom by means of a yoke weldment. The upper one or two stages were controlled by a single internal cylinder. A four-stage two-leg gantry of this style is shown in Fig. 1.13 undergoing a load test at the Lift Systems factory.

Figure 1.12 J&R Engineering Telescopic Boom Gantry *(Burkhalter Rigging, Inc.)*

A second feature of the Lift Systems telescopic boom gantries was the use of boom sections fabricated from four plates welded together to form a rectangular box section, rather than the then-dominant structural tube boom sections. This practice allows the use of high-strength alloy steels for boom construction as well as a "fine tuning" of the proportions of the boom sections for structural efficiency.

Figure 1.13 Four-Stage Lift Systems Telescopic Boom Gantry *(Lift Systems, Inc.)*

Boom design developments advanced by J&R Engineering beginning in 1999 included the use of boom sections with an octagonal cross section. These boom sections are fabricated using cold-formed steel plates and, like the four-plate sections, can be proportioned to match the boom section strength to the particular demands of the leg design.

Later versions of the Lift Systems telescopic boom gantries were developed using one internal cylinder per leg (Fig. 1.14). Some gantries so configured included a manual boom section and others had all boom sections powered. Other enhancements included pin locks that mechanically lock boom sections together at specific intervals of extension and wedge locks that lock boom sections together at any extension position.

The first commercial manufacturer of hydraulic gantries outside of the United States was Hydrospex Cylap B.V., a Netherlands-based company then owned and managed by Tjerko Jurgens. Hydrospex began manufacturing bare cylinder gantries in 1994. The original Hydrospex gantry model (in the foreground in Fig. 1.15) was a simple bare cylinder leg with an external control and power unit. The design of this gantry model was later modified to incorporate an internal hydraulic oil tank, motor, and pump, thus making each gantry leg a self-contained unit (Fig. 1.16). This basic gantry leg configuration has remained a constant within the Hydrospex product line to date.

Hydrospex launched the SBL line of telescopic boom gantries (Fig. 1.17) in 2006. This model line introduced two innovative features. These are the use of a roller chain, rather than wheels, to provide each leg with a larger footprint and a base design that allows the boom to be pivoted down to a horizontal position for transportation in a standard 20-foot container.

Figure 1.14 Four-Leg Gantry System Lifting a Generator *(Bigge Crane and Rigging Co.)*

Figure 1.15 Original Hydrospex Bare Cylinder Gantry *(Enerpac Integrated Solutions)*

A second European company, Greiner GmbH–Fahrzeugtechnik of Germany, entered the gantry business in the late 1990s. Greiner, founded in 1980 by Karl and Margot Greiner, began as a design office and later introduced a line of hydraulic

Figure 1.16 Hydrospex Bare Cylinder Gantry *(Enerpac Integrated Solutions)*

Figure 1.17 Hydrospex Telescopic Boom Gantry *(Enerpac Integrated Solutions)*

platform trailers. The company started the manufacture of bare cylinder hydraulic gantry systems in 1999. Greiner has remained focused on bare cylinder gantries, with systems ranging from simple single stage legs (Fig. 1.18) to large, three-stage systems (Fig. 1.19).

Figure 1.18 Greiner Single Stage Bare Cylinder Gantry *(Greiner GmbH)*

Figure 1.19 Greiner Three-Stage Bare Cylinder Gantry *(Greiner GmbH)*

Electronic operator aids became available for gantry systems almost as far back as the birth of the commercial gantry industry. Lift Systems began marketing in 1986 a device by which the level of each header beam could be monitored and J&R Engineering began offering in the early 1990s attachments that monitored the header beam level and later introduced an attachment that provided a readout of the extension height of each gantry leg. Load indicators also became available from all three of the U.S. gantry manufacturers in the mid-1990s. Initially hard wired, these systems were soon enhanced to include wireless readout units. These devices only provided the operator with information about the movements of the system or the loads carried at each leg.

Fully computerized gantry system controls would not come along until the early 2000s. In 2002, Hydrospex advanced the use of electronic controls with the introduction of it's Intellilift control system. This comprehensive computer-based system monitors the supported load, leg extension and travel distance to provide the operator with feedback, has the ability to set limits to leg functions, such as overload detection, and can automatically synchronize the various functions of the gantry system. In 2005, 4-Point Lift Systems, Inc. and Riggers Manufacturing Company began providing the similarly comprehensive CARL (Computer Assisted Remote Lifting) control system across their gantry lines. Greiner began integrating computer-monitoring and remote controls into their gantry systems around 2007. J&R Engineering's introduction of the Lift Equalizer wireless control system rounded out the computer control offerings from the major gantry manufacturers.

In 1996, ownership of the first of the current gantry manufacturers, Riggers Manufacturing Company, changed. The company was purchased by David J. Pokraka. This change in ownership did not change either the company's location or its product line.

In 2002, the gantry business changed once more. Ownership of Riggers Manufacturing Company was purchased by a joint venture of 4-Point Lift Systems, Inc. and Rigging Gear Sales, Inc., of Dixon, Illinois. The Riggers Manufacturing product line again remained intact after this change in ownership.

Three years later, in 2005, 4-Point Lift Systems, Inc. was sold to a group led by Bruce Forster, the owner of Rigging Gear Sales, Inc., and the company name was formally shortened to Lift Systems, Inc. This purchase gave Forster a controlling interest in two of the three U.S. gantry manufacturers.

Hydrospex changed ownership in 2010 when the company was purchased by Actuant Corporation, a company based in the United States that also owns hydraulic jack manufacturers Enerpac and Simplex. Hydrospex products are sold and serviced through the company's Enerpac Integrated Solutions unit. By 2012, the Hydrospex name had been retired and the hydraulic gantry products are now being sold under the Enerpac name.

Today, telescopic hydraulic gantry systems are available in a wide variety of capacities ranging from 10 tons (9.1 tonnes) to 345 tons (313 tonnes) per leg and systems with greater capacities are in the design pipeline. Multiple stage cylinders and booms are used on the vast majority of the systems manufactured today and electronic control systems enhance safety and ease of use. Just as this type of equipment has grown from one-of-a-kind machines and specialty products in the 1960s to the broad array of sophisticated gantry systems we see today, there is every reason to expect that the products will continue to evolve in the years to come as user demand spurs new design efforts and the manufacturers continue to take advantage of advancing technology.

1.4   INDUSTRY GUIDELINES AND STANDARDS

The development of hydraulic gantry systems and their use progressed with minimal formal guidance from industry groups and no regulatory control from either government agencies or standard-writing organizations. It was not until the 1990s that the industry felt the need to begin development of standards that address hydraulic gantries.

The first effort by the industry to develop guidelines for the use of hydraulic gantries began in 1994 when the Specialized Carriers & Rigging Association (SC&RA), a construction industry trade group based in Centreville, Virginia, convened the Jacking Systems Task Force. This effort was made in response to a number of significant accidents that occurred during the use of hydraulic gantries in industrial construction work. The task force was charged with developing a set of basic guidelines for the use of hydraulic gantry systems. This task force produced *Recommended Practices for Hydraulic Jacking Systems* (SC&RA 1996), which was published by SC&RA in 1996.

In response to changes in the industry, the SC&RA Telescopic Hydraulic Gantry Task Force was convened in early 2004 and was charged with updating and expanding the provisions of the earlier document to meet the industry's current needs. The final product of this task force was *Recommended Practices for Telescopic Hydraulic Gantry Systems* (SC&RA 2004), which was published at the end of 2004.

Shortly after completing this publication, SC&RA produced a training video with the same title. This video, developed with the support of the three U.S. gantry manufacturers, serves as a basic employee training tool for contractors engaged in the use of hydraulic gantry systems.

The U.S. Occupational Safety and Health Administration (OSHA) informally participated in the development of the SC&RA *Recommended Practices for Telescopic Hydraulic Gantry Systems*. Although this publication is only a guide and not a formal standard, the industry anticipates that OSHA will use this document as a basis for job site inspections and accident investigations. There are no OSHA regulations at present that directly address the use of hydraulic gantries. [Note that the OSHA regulations published in 29 CFR 1910.244 and 29 CFR 1926.305 apply to hydraulic jacks of the type illustrated in Fig. 1.20 and are not applicable to hydraulic gantry systems. Further, telescopic hydraulic gantries are specifically excluded from the scope of OSHA's cranes and derricks regulations (29 CFR 1926.1400 (c)(6).]

Safety standards for many types of lifting equipment are written in the U.S. by the American Society of Mechanical Engineers (ASME). Standards that address mobile and overhead cranes, derricks, and related lifting equipment are grouped under the B30 designation. In 1998 and 1999, ASME explored the development of a B30 volume for hydraulic gantry systems. The idea did not proceed due to a lack of demand on the part of both the manufacturers and the users of hydraulic gantries. The prospect was revisited in 2004, this time with the proposal of adding a chapter on gantry systems to the existing volume B30.1-2004 *Jacks* (ASME 2004), which at the time addressed only basic hydraulic and mechanical jacks. This proposal

Figure 1.20 Basic Hydraulic Jack

was met with support from both users and manufacturers of hydraulic gantries. The B30.1 volume has been expanded to add coverage for a number of types of rigging gear, including telescopic hydraulic gantry systems. The volume title is now B30.1-2009 *Jacks, Industrial Rollers, Air Casters, and Hydraulic Gantries*. Chapter 6 addresses the use of telescopic hydraulic gantry systems.

The Infrastructure Health & Safety Association, formerly the Construction Safety Association of Ontario, initiated a project in 1997 to develop a safety program for lifting with hydraulic gantries. The result of this work is a chapter in *Specialized Rigging – Safety Guidelines for Construction* (IHSA 2003). The guide covers a variety of types of equipment, including hydraulic gantry systems. The format of the chapter on hydraulic gantries is very similar to that of the first SC&RA guide (SC&RA 1996), but with a greater emphasis on field work and without detailed engineering guidance. This publication has been replaced by the *Construction Multi-Trades Health and Safety Manual* (IHSA 2007). The gantry material appears in Chapter 25, unchanged from that in the older publication.

The U.K. Health and Safety Executive (HSE) sponsored a study begun in 1997 on the use of hydraulic gantry systems. This project was prompted in great part by a major industrial accident with a hydraulic gantry system that occurred in the U.K. in 1994. AEA Technology plc, a consulting firm based in Harwell, Oxfordshire, U.K. conducted an extensive survey of hydraulic gantry system users and manufacturers worldwide to compile data on the design, manufacture, and use of gantries. This information was assembled into a series of reports that was used by the British Standards Institution (BSI) in the development of a safety standard for hydraulic gantries for use in the U.K. A draft of the standard (BS 7121-13 *Code of Practice for Safe Use of Cranes – Part 13: Hydraulic Gantry Lifting Systems*) was released by BSI in September 2002 for public comments. The first edition of the standard was published in 2009 (BSI 2009).

The provisions of the current standards and how they affect lift planning and engineering are discussed in Chapters 6 and 7.

1.5   ADDITIONAL READING

Most of the companies discussed in this chapter are or were members of the Specialized Carriers & Rigging Association (formerly the Heavy Specialized Carriers Conference). Additional information about these companies, their products, and their projects can be found in the Association's publications. Following in chronological order is a brief list of articles that may be of interest to the reader. (The association name and location shown were current at the time of each article's publication.)

"Moving 400-ton Loads 65 Miles Over Virginia's Secondary Roads," *Transportation Engineer*, Vol. 20, No. 11, July 1972, Heavy Specialized Carriers Conference, Washington, D.C.

"Linden's Big Job: Tight Quarters," *Transportation Engineer*, Vol. 29, No. 6, July 1982, Specialized Carriers & Rigging Association, Washington, D.C.

"Lift Systems Pioneers Adaptable Lifting Device," *Transportation Engineer*, Vol. 31, No. 8, September 1984, Specialized Carriers & Rigging Association, Washington, D.C.

"Chip Belding Back Into Fold at Family-Run Belding Corp.," *Transportation Engineer*, Vol. 31, No. 11, December 1984, Specialized Carriers & Rigging Association, Washington, D.C.

"Riggers Telescoping E-Z LIFT System Owes Debt to Hydraulic Gantry Development," *Transportation Engineer*, Vol. 32, No. 3, April/May 1985, Specialized Carriers & Rigging Association, Alexandria, VA.

"Hydratech Offers Three Distinct Gantries," *Transportation Engineer*, Vol. 32, No. 3, April/May 1985, Specialized Carriers & Rigging Association, Alexandria, VA.

"Expert Gantry User," *Transportation Engineer*, Vol. 33, No. 1, January 1986, Specialized Carriers & Rigging Association, Alexandria, VA.

"With Riggers In Mind," *Transportation Engineer*, Vol. 33, No. 3, April/May 1986, Specialized Carriers & Rigging Association, Alexandria, VA.

"The 'Riggers' of Work," *Lifting & Transportation International*, Vol. 39, No. 7, September 1992, Specialized Carriers & Rigging Association, Fairfax, VA.

"Alternative Lift Equipment," *Lifting & Transportation International*, Vol. 42, No. 4, May 1995, Specialized Carriers & Rigging Association, Fairfax, VA.

"Riggers Manufacturing Sold," *Lifting & Transportation International*, Vol. 43, No. 4, May 1996, Specialized Carriers & Rigging Association, Fairfax, VA.

"Lift Systems Joint Venture Acquires Riggers," *Lifting & Transportation International*, Vol. 49, No. 6, December 2002/January 2003, Specialized Carriers & Rigging Association, Fairfax, VA.

Hoffman, P.E., Norman, and Hale, P.E., William, "Classification Dilemma," *International Cranes and Specialized Transport*, Vol. 12, No. 4, January 2004, Specialized Carriers and Rigging Association, Fairfax, VA.

"Super Beam Lift," *International Cranes and Specialized Transport*, Vol. 13, No. 10, July 2005, Specialized Carriers and Rigging Association, Fairfax, VA.

"Specialist Consortium Acquires Lift Systems," *American Cranes & Transport*, Vol. 1, No. 3, August 2005, Specialized Carriers & Rigging Association, Fairfax, VA.

"Gantry Founder," *American Cranes & Transport*, Vol. 1, No. 4, September 2005, Specialized Carriers & Rigging Association, Fairfax, VA.

"Old Tool, New Uses," *American Cranes & Transport*, Vol. 2, No. 9, September 2006, Specialized Carriers & Rigging Association, Fairfax, VA.

"SC&RA 2009 Jobs of the Year," *American Cranes & Transport*, Vol. 5, No. 6, June 2009, Specialized Carriers & Rigging Association, Fairfax, VA. [Note: Two of the three Rigging Job of the Year winners in 2009 utilized hydraulic gantry systems.]

"Hydrospex Acquired by Actuant," *International Cranes and Specialized Transport*, Vol. 18, No. 8, May 2010, Specialized Carriers and Rigging Association, Fairfax, VA.

"SC&RA Jobs of the Year," *American Cranes & Transport*, Vol. 8, No. 5, May 2012, Specialized Carriers & Rigging Association, Fairfax, VA. [Note: All three Rigging Job of the Year winners in 2012 utilized hydraulic gantry systems.]

1.6 REFERENCES

American Society of Mechanical Engineers (ASME) (2004), ASME B30.1-2004 *Jacks*, New York, NY.

American Society of Mechanical Engineers (ASME) (2009), ASME B30.1-2009 *Jacks, Industrial Rollers, Air Casters, and Hydraulic Gantries*, New York, NY.

British Standards Institution (BSI) (2009), BS 7121-13:2009 *Code of Practice for Safe Use of Cranes – Part 13: Hydraulic Gantry Lifting Systems*, London, U.K.

Infrastructure Health & Safety Association (IHSA) (2003), *Specialized Rigging – Safety Guidelines for Construction*, Mississauga, Ontario.

Infrastructure Health & Safety Association (IHSA) (2007), M033 *Construction Multi-Trades Health and Safety Manual*, Mississauga, Ontario.

Specialized Carriers & Rigging Association (SC&RA) (1996), *Recommended Practices for Hydraulic Jacking Systems*, Centreville, VA.

Specialized Carriers & Rigging Association (SC&RA) (2004), *Recommended Practices for Telescopic Hydraulic Gantry Systems*, Centreville, VA.

2 Gantry System Basics

Telescopic hydraulic gantry systems serve the rather basic function of lifting a load through a relatively small height range. Lifting capacities of commercial gantry systems with four telescopic legs currently range from 40 tons (36 tonnes) to 1,380 tons (1,252 tonnes). The fully extended heights of many of the more common gantry models are in the 25- to 30-foot (7.5- to 9-meter) range, and some models have heights ranging up to 48 feet (14.6 meters). The gantry legs typically are equipped with wheels or rollers that allow the system to travel while supporting a load. Some gantry systems are also equipped with devices to allow horizontal movement of the suspended load in the direction perpendicular to the travel direction.

In this chapter, the various types of gantry products that are currently in use are reviewed. It is emphasized that the some of the gantry system applications shown here are specialized arrangements that may not be suitable for all lifting situations. The user of a gantry system must understand fully the use and operation of the equipment and the conditions on the job site.

2.1 GANTRY SYSTEM COMPONENTS AND FEATURES

Hydraulic gantry systems share many of the same components and features. The following sections provide brief descriptions of the most common elements of gantry systems as currently used.

2.1.1 The Gantry Leg

The fundamental component of the gantry system is the gantry leg (Fig. 2.1). There are two basic styles of gantry leg. These are the telescopic boom leg and the bare cylinder leg.

The telescopic boom gantry leg consists of a box-like base weldment, usually fabricated from steel plate, with a tubular steel multiple section lift boom, similar in appearance to the boom of a hydraulic crane. The design of this leg is such that the hydraulic cylinder(s) that extend and retract the boom support only the vertical load and any horizontal loads that act on the leg are resisted by the lift boom. An important advantage of some models of the telescopic boom gantry leg is that the

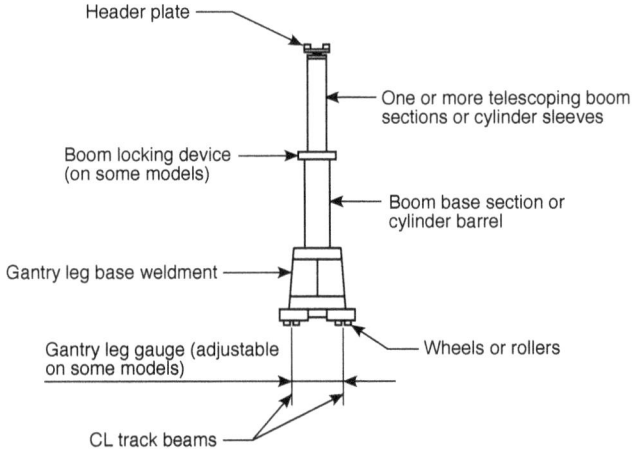

Header plate

One or more telescoping boom sections or cylinder sleeves

Boom locking device (on some models)

Boom base section or cylinder barrel

Gantry leg base weldment

Gantry leg gauge (adjustable on some models)

Wheels or rollers

CL track beams

Figure 2.1 Basic Telescopic Hydraulic Gantry Leg

boom sections can be locked together mechanically using either pins or gripping devices so that the load can be supported structurally, rather than hydraulically. This feature may be desirable if a load is to be lifted and held by the gantry system for an extended length of time.

The bare cylinder gantry leg consists of a similar base weldment with one or more vertical hydraulic lift cylinders that serve to raise and lower the load. In addition to providing all vertical support, the exposed lift cylinders must also resist any horizontal loads to which the leg is subjected. A bare cylinder gantry is much less complex mechanically and, therefore, is typically lighter to transport and handle and easier to maintain, as compared to a telescopic boom gantry of similar lift height and rated load.

Fig. 2.1 illustrates the simplest form of the gantry leg and identifies the most significant components of the leg. In the form shown, the leg has one lift boom or lift cylinder and in the case of the boom type of leg, the actuating cylinder is enclosed within the boom. Other bare cylinder gantry leg models are or have been manufactured with two, three, or four lift cylinders (Fig. 2.2). Some telescopic boom gantry legs have two cylinders on either side of the lift boom, rather than a single cylinder within the boom (Fig. 2.3). The header plate at the top of the lift cylinder or lift boom provides a positive connection point for attachment of the header beam. The header plate is typically attached to the cylinder or boom by a pinned connection, a rocker fixture, or a ball and socket joint.

Most gantry models are constructed with fixed-position wheels mounted to the underside of the base weldment (Fig. 2.4a). Both the wheelbase and the gauge of the gantry leg are fixed with this style of wheel arrangement. One gantry system manufacturer produces some gantry leg models on which the wheels are mounted in bolt-on wheel boxes (Fig. 2.4b). Although the wheelbase is still fixed, the gauge of the gantry leg is adjustable. This allows the user to position the wheels at the wider gauge for greater stability when working at higher leg extensions or

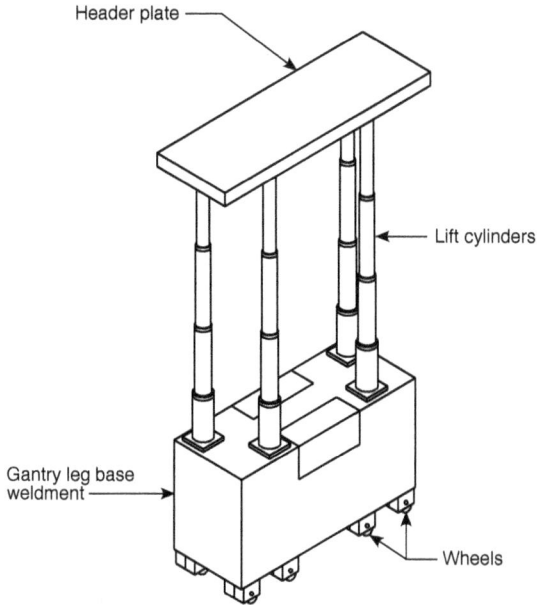

Figure 2.2 Four-Cylinder Gantry Leg

to mount the wheels in the retracted position to reduce the width of the leg when working in locations with very limited space. A third variation of leg design makes use of two roller chains, one over each track beam, rather than individual wheels

Figure 2.3 Telescopic Boom Gantry with External Cylinders *(Lift Systems, Inc.)*

(a)   (b)   (c)

Figure 2.4 *(a)* Fixed Wheels. *(b)* Adjustable Wheel Boxes. *(c)* Rollers.
(David Duerr, P.E.; Barnhart Crane & Rigging Company)

(Fig. 2.4c). These roller chains are similar in appearance and function to those found in industrial rollers and provide a long footprint on each track surface, rather than a few discrete point loads.

Some gantry leg models are outfitted with control system data acquisition devices such as height indicators that measure the vertical stroke of the gantry leg, travel sensors that measure the movement of the leg along the track, and load sensors, such as hydraulic fluid pressure transducers or instrumented shear pins, that measure the weight being supported by the leg. These devices typically send electrical signals to the operator's control unit, but on some models, the output from these devices must be read at the individual gantry legs. The functions of these data acquisition devices within the overall gantry control system are discussed in Section 2.1.4.

2.1.2 Gantry System General Arrangements

Fig. 2.5 illustrates the primary components of a basic two-leg gantry system. The track beams are typically structural members, most often box shaped, that provide a smooth and guided surface along which the system may travel. The tracks may also serve to span between hard supports to carry the relatively high gantry loads over an opening or across a lightweight floor structure. The gantry leg, discussed in the preceding section, is the manufactured product that is the heart of the system. Spanning between the two gantry legs is the header beam, which is typically a rolled wide flange shape or a fabricated box-shape beam. The lift links, sometimes referred to as D-rings, are fittings or devices that provide a means for attaching the rigging to the header beam. The locations of the lift links on the header beam are called the load points.

The beam arrangement used for a lift may vary from one simple header beam spanning between two legs, as shown in Fig. 2.5, to complex arrangements that provide rigging connections to multiple lift points on the load. One common configuration used to provide increased flexibility in the lift point geometry is

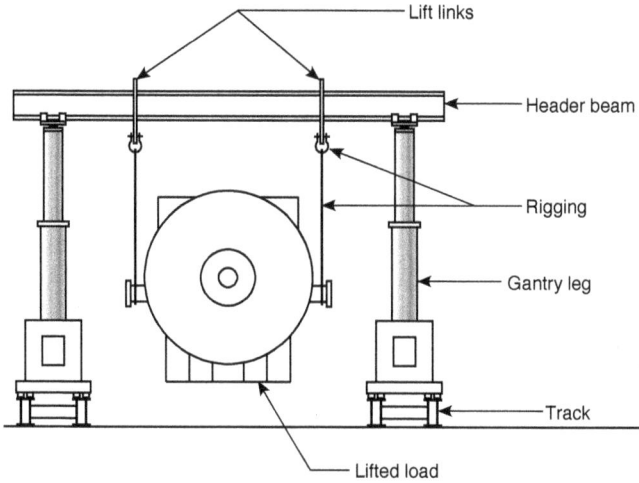

Figure 2.5 Two-Leg Hydraulic Gantry System

the four-beam arrangement, an example of which is illustrated in Fig. 2.6. The beams mounted directly on the gantry legs are called header beams (that is, the header beams attach to the gantry legs' header plates) and the upper beams that run crosswise to the header beams are called cross beams. This four-beam setup offers great flexibility in that it allows the gantry legs and track to be located to suit available clearances and support conditions while independently allowing the

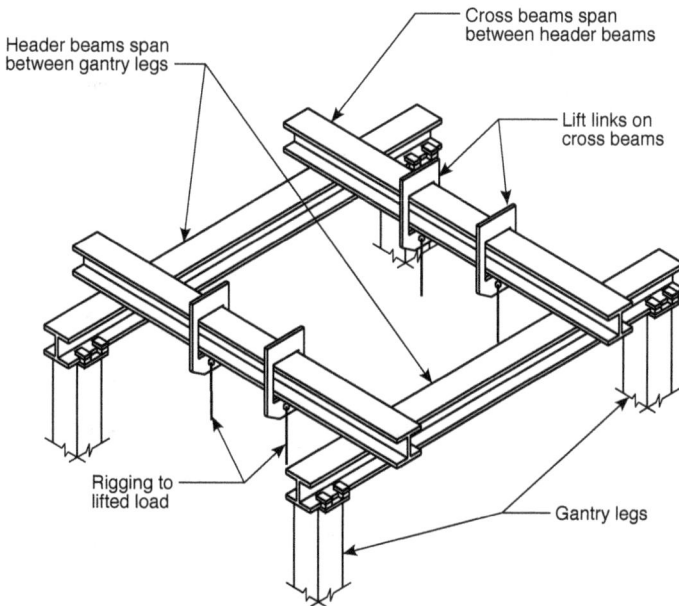

Figure 2.6 Four-Beam Arrangement

locations of the load points on the cross beams to correspond to the positions of the lifting attachments (lifting lugs, trunnions, etc.) on the load.

The beam arrangement illustrated in Fig. 2.6 is the simplest form of the four-beam setup in that the layout is symmetrical and the cross beams are perpendicular to the header beams. As an example of the flexibility offered by the four-beam arrangement, consider the lift shown in Fig. 2.7. The header beams are parallel to one another and perpendicular to the direction of travel. The cross beams are angled in order to spread the forward slings at an outward angle to clear external components on the lifted load. When using an arrangement of this sort, it is vital that sling tensions are accurately calculated and that the cross beams are adequately secured to the header beams so as not to shift under the influence of the horizontal force components of the sling tensions.

The general orientation of the header and cross beams may be chosen to suit the requirements of the lift. That is, the header beams may be oriented perpendicular to the direction of travel, as shown in Fig. 2.7, or they may be parallel to the direction of travel, as shown in Fig. 2.8. In either case, the cross beams span between the header beams.

Associated with the header and cross beams are the lift links that provide the means of attaching the rigging to the beams. There is a variety of lift link styles available today. The basic lift link is simply a flat plate with a rectangular hole through which the header or cross beam passes and one or more round holes below that into which a shackle or other pin-connected device is fitted, as seen in Figs. 2.5 through 2.8. This is the most common type of lift link due to its simplicity and relatively light weight. More sophisticated lift links are available that are equipped with powered sliding or rolling devices that allow the load to be moved along the length of the beam (a movement called side shifting). A set of powered side shifting lift links is shown in Fig. 2.9. Conventional lift links can be adapted for

Figure 2.7 Four-Beam Lift with Angled Cross Beams *(Taylor Crane & Rigging, Inc.)*

Figure 2.8 Header Beams Parallel to Travel Direction *(Lift Systems, Inc.)*

side shifting by means of dollies. A pair of adjustable height lift links mounted on side shift dollies is shown in Fig. 2.10. Given the wide range of lift link products that are available, details of the various types of lift links are best obtained from the literature or web sites of the gantry system manufacturers. As with most gantry system accessories, some rigging contractors design and manufacture their own lift link devices to match their unique needs.

The selection of the appropriate type of lift link for a particular project is purely a function of the demands of the lift. The majority of gantry lifts are of the rig-it-lift-and-travel type for which the plate style lift links are suitable. Other types of lift links may be used where the circumstances of the lift require side shifting or rotation of the suspended load.

Figure 2.9 Powered Lift Links *(J&R Engineering Co., Inc.)*

Figure 2.10 Adjustable Lift Links with Side Shift Dollies *(David Duerr, P.E.)*

2.1.3 Gantry System Track and Propel Devices

All of the gantry system manufacturers produce track assemblies that are designed to be used with their gantry products. A basic track assembly is shown in Fig. 2.11. Rigging contractors most often use the track available from the manufacturer for their particular gantry systems. This is not to imply that these track assemblies are appropriate for all applications, however, or that it is inappropriate for a contractor to design and fabricate its own track. A number of areas must be investigated to assure that the selected track is acceptable and for some lifts, specially designed track beams are needed. Track design is treated in detail in Chapter 6.

The main features of the track assembly are the track beams along which the gantry legs move, one or more wheel guide bars to keep the gantry wheels centered on the beams (Fig. 2.12), cross members to hold the beams in position, and the propel cylinder anchor when required, as discussed later.

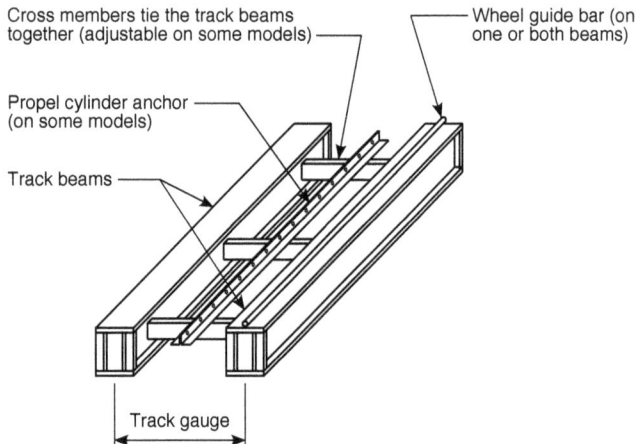

Figure 2.11 Gantry Track Assembly

Figure 2.12 View of Gantry Wheels and Wheel Guide Bar *(David Duerr, P.E.)*

Some lifts require that the track beams be raised to an elevation that cannot be achieved practically using timber cribbing. This can be accomplished by means of structural support frames commonly referred to as track stands (Fig. 2.13). Some of the gantry system manufacturers produce track stands that are compatible with their track assemblies. Track stands are also designed and fabricated by the contractors that use gantry systems to suit the particular demands of their projects. Regardless of the source, the track stand strength design must be consistent with that used for track design to provide adequate strength and stability. Depending on the height, the track stands may have to be braced in one or both horizontal directions to assure adequate stability.

Travel of the gantry system along the track is accomplished by means of a propel system. There are three types of propel systems presently in common use.

Figure 2.13 Gantry System Supported on Track Stands *(Lift Systems, Inc.)*

These are the built-in (or integral) drive, the external drive wheel, and the external propel cylinder.

The built-in, or integral, drive system is, as the term implies, internal to the gantry leg. Typically, the system consists of a hydrostatic drive motor and a gear train or a chain drive that connects the motor to some or all of the gantry leg's wheels or rollers. In most designs, the drive can be disengaged to allow the leg to freewheel. (The value of the ability of the gantry leg to freewheel is discussed in Chapter 7.) Use of the integral drive is not practical on gantry legs that have adjustable wheel boxes (Fig. 2.4b).

The external drive wheel (Fig. 2.14) is a unit consisting of a hydrostatic motor, gearing or a chain drive, and a wheel with a high-friction tread surface. The unit is attached to the leg such that the wheel contacts the track surface. The wheel is preloaded to bear on the track by means of an adjustable compression link, thus providing drive traction. Two drive wheels per leg, one at each track, are necessary to provide a balanced drive force.

The external propel cylinder system (Fig. 2.15) utilizes a hydraulic cylinder installed between the leg and the track to push or pull the gantry system. Application of this propel system requires the use of track assemblies that provide anchor points for the cylinder. The most common arrangement uses a pair of angles that are located on the centerline of the track assembly and welded to the assembly cross members (Fig. 2.11). The stationary end of the cylinder is pinned to sets of holes in the vertical legs of the angles. Propel cylinders are most commonly off-the-shelf models. However, if special cylinders are designed, they must be proportioned for adequate capacity in the retract mode and for adequate buckling strength in the extend mode.

An arrangement in which one gantry leg on the track is driven and the other is not and where the legs must move together requires the use of one or two connecting members (Fig. 2.15). These members and their attachments to the

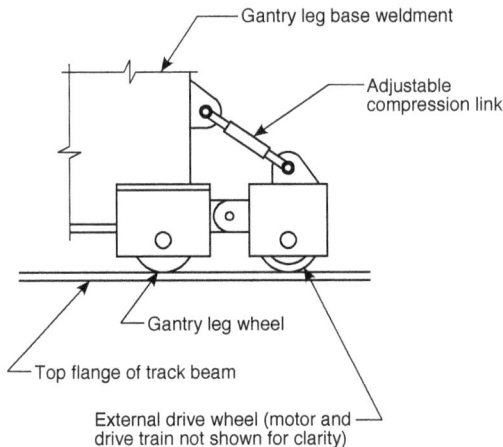

Figure 2.14 Gantry Leg External Drive Wheel Propel System

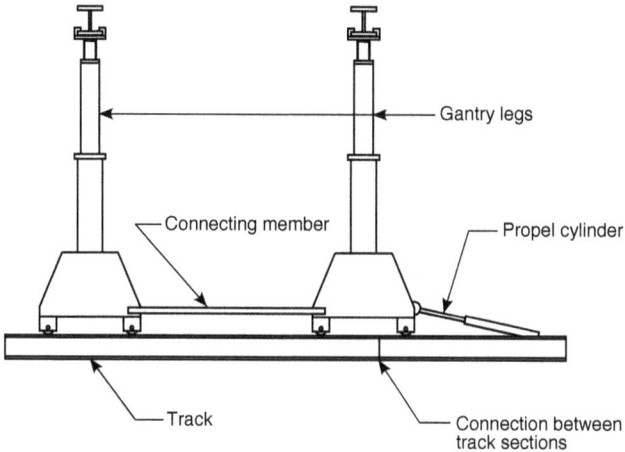

Figure 2.15 Gantry External Cylinder Propel System

gantry legs must be strong enough to transmit the drive force from the driving leg to the driven leg. The connecting members should be capable of carrying the drive force in either tension or compression in order to provide full control of the driven leg. The general recommendation for the layout of this type of propel system calls for attaching the propel cylinders to the forward legs such that the connecting members are in tension while traveling. The ability to carry the drive force in compression recognizes that the need may exist to reverse direction as well as to provide braking control.

2.1.4   Control and Power Unit

The last major component of a gantry system is the control and power unit. The current range of gantry systems on the market employ a variety of control and power unit configurations.

Some gantry systems utilize a control and power unit that is a stand-alone assembly containing the oil tank, hydraulic pump, a gasoline, diesel, or propane engine or an electric motor that drives the pump, and the various switches, valves, gauges, and other controls that provide the gantry operator with the means to operate the system (Fig. 2.16*a*). This stand-alone control and power unit is connected to the individual gantry legs by sets of hydraulic hoses. This arrangement is referred to as an all-hydraulic system. Other gantry models have an oil tank, hydraulic pump, and engine/motor (collectively, the power unit) housed in one assembly (Fig. 2.16*b*), again connected to the gantry legs by hoses, and an electric control station (the control unit) connected to the system by cables or by radio. This arrangement is referred to as an electric-over-hydraulic system. A third arrangement in common use places a power unit within the base of each gantry leg (Fig. 2.16*c*) and has a cable- or radio-connected electric control unit to operate the system. One model of remote control unit is shown in Fig. 2.16*d*. Many of the remote control units can operate wirelessly or they can be hard-wired to the gantry system.

Gantry system control units are increasingly incorporating electronic devices to monitor the load supported by each leg, the extension height of each leg, and the travel and side shift motions. These operator aids, much like the computer systems that are now widely used on mobile cranes, provide a great deal of feedback to the operator and can assist in the operation of the gantry system.

These electronic control system features require the installation of sensors on the gantry legs that can measure the quantities of interest (e.g., load, wheel rotation, etc.) and a suitably programmed computer within the control unit that can process the measurements and convert these data into a form usable by the operator, such as information displayed on a screen. The most sophisticated systems include logic routines that utilize the information from the sensors on the gantry legs to automatically synchronize the motions and control the accelerations, both vertical and horizontal, of the legs during the performance of a lift.

2.2   BASIC GANTRY SYSTEM APPLICATIONS

Hydraulic gantries are used by specialty rigging and transportation contractors for a wide variety of lifting operations. In the most general sense, the use of a hydraulic gantry system for heavy lifting is not fundamentally different than the

(a)

(b)

(c)

(d)

Figure 2.16 *(a)* External Control and Power Unit. *(b)* External Power Unit. *(c)* Internal Power Unit. *(d)* Remote Control Unit. *(David Duerr, P.E.)*

use of a mobile or overhead crane. The manufacturer provides a product with a set of operating instructions and a load chart. The user is then responsible for the selection of the rigging, the preparation of the base on which the equipment will be set up, and operation of the equipment in a safe manner and in accordance with the manufacturer's operating instructions.

The hydraulic gantry system was originally conceived for lifting and moving heavy machinery components inside buildings where clearances are slight, mobile cranes are not practical, and overhead cranes of adequate capacity are not available. Over the years, however, gantry systems have found their way onto construction sites as an efficient and economical alternative to mobile cranes for lifting heavy loads. Despite the wide range of lifting environments in which hydraulic gantries are used today, the types of lifts for which gantry systems are most often used can be summarized in five categories.

2.2.1   Lifting with No Horizontal Movement

The simplest gantry lifting operation calls for only vertical movement of the load. The item to be lifted is rigged to the header beams or cross beams, the gantry legs are extended to lift the load, and then retracted to lower the load.

A common application of this type of lift is the unloading of an item from a vehicle and setting it down onto storage cribbing or onto another vehicle (for example, the transfer of a load from a rail car to a heavy-haul trailer). This type of lift is performed by setting up the gantry system and then bringing the laden vehicle into position between the gantry legs. (The reverse of this process may also be done; that is, the gantry system is erected over the stationary vehicle and load.) An example of this type of lift is illustrated in Fig. 2.17. The load is rigged to the header or cross beams and the gantry legs are extended to lift the load free of the delivery vehicle (Fig. 2.17a). The delivery vehicle is then moved out, storage

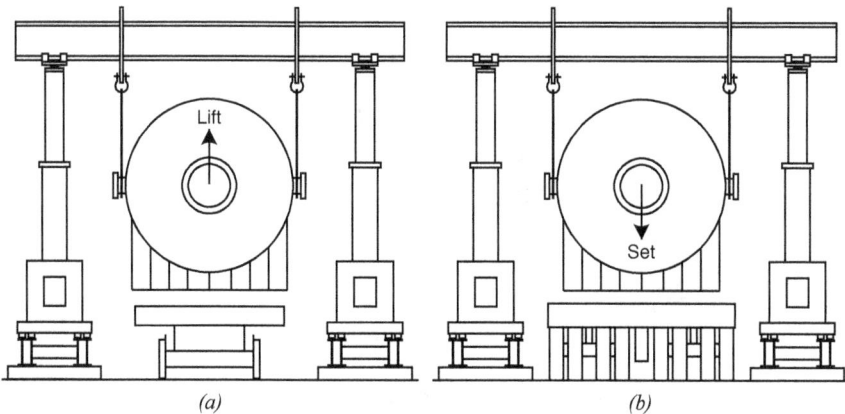

(a)   *(b)*

Figure 2.17 Basic Hydraulic Gantry Lifting Operation

cribbing or the receiving vehicle is positioned below the load, and the load is then lowered to its final position by retracting the legs (Fig. 2.17b). When storage cribbing is to be used, the work must be planned to minimize the positioning of workers under the suspended load.

2.2.2 Lifting with System Travel

This lifting operation calls for lifting the load, traveling the gantry system along track, and then setting the load. This method is commonly used for machinery installation, an example of which is shown in Fig. 2.18. A machinery component, here a generator, is delivered to a point adjacent to one end of its foundation. It is lifted with the gantry system, traveled to a position over its anchor bolts, and then lowered into its rough-set position.

The most important lift planning issues that must be addressed for a "lift and travel" operation are selection of a track system that can accommodate the necessary gantry movement and provision of supports for that track system that are proportioned to carry the maximum loads that occur as the gantry system travels. Additional considerations include evaluation of vertical and horizontal clearances for the load and the gantry system over the full extent of travel.

2.2.3 Lifting with Side Shifting

In a variation of the lift-and-travel operation, the gantry system may be configured to move the suspended load in the lateral direction (perpendicular to the direction of travel), a movement called side shifting (Fig. 2.19). Again using the example of moving a load from a rail car to a trailer, application of this technique calls for a gantry system that spans both the rail car and the trailer. The load is rigged to

Figure 2.18 Lift and Travel to Set a Generator *(J&R Engineering Co., Inc.)*

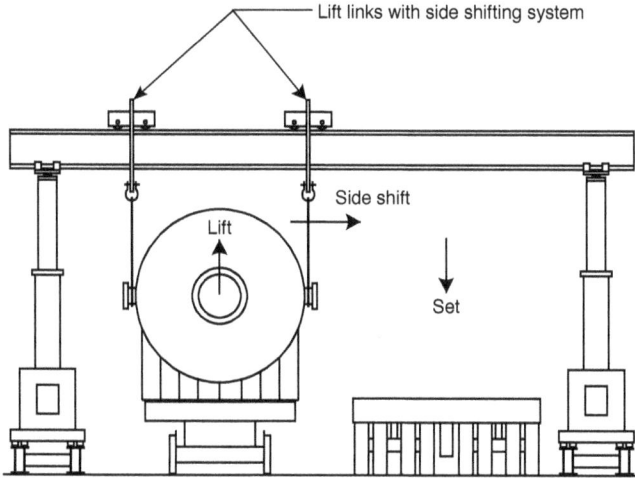

Figure 2.19 Side Shifting a Load from Rail Car to Trailer

the header beams using lift links with a side shifting capability, lifted from the rail car by extending the gantry legs, moved laterally along the header beams using the side shifting system, and then lowered to the trailer by retracting the gantry legs. (This is the type of lift seen in Fig. 2.9, although the rail car was not in the area at the time the photograph was taken.)

This method is also used to lift a load from a vehicle and set it on a foundation. In this type of operation, the load is brought into position along side the foundation, rather than to the end, and the gantry system is set up to straddle the trailer and the foundation. The load is then lifted from the trailer, side shifted to a position over its anchor bolts, and then lowered into its rough-set position. Setting up the gantry system on track, as seen in Figs. 2.9 and 2.19, enhances the versatility of this arrangement by allowing easy movement in both horizontal directions when placing the load.

If the lifted load is relatively light, side shifting may be facilitated though the use of an underhung trolley running on the bottom flange of the header beam (Fig. 2.20). The header beam design for this application must address the local stresses in the bottom flange that are developed by the concentrated loads from the trolley wheels. An appropriate beam design method for this application is presented in detail in Chapter 6. (And in answer to the obvious question, yes, that is the real Liberty Bell shown being lifted in Fig. 2.20.)

2.2.4   Upending or Laying Over

Hydraulic gantry systems are also used to upend or lay over (downend) tall items (Fig. 2.21). This application is commonly driven by the need to install or remove pieces of equipment that are shipped in a horizontal orientation, but must

Figure 2.20 Use of an Underhung Trolley *(J&R Engineering Co., Inc.)*

be installed in the vertical position. This is a more complex operation than the straight lifting described in the previous sections because, in the most common application of this type of lift, the coordination of simultaneous horizontal and vertical movement of the two pairs of gantry legs is required. Further, the load distribution between the two pairs of legs will often change as the load is rotated from horizontal to vertical.

In such an upending (or downending) operation, the gantry legs supporting the upper end of the load (the lifting gantry) are typically stationary and lift (or lower) only, while the gantry legs controlling the lower end (the tailing gantry) are rolled along the tracks to bring the bottom of the load under the top. Such an operation is functionally similar to upending or downending a load using two mobile cranes, with one crane lifting the top of the load and the other tailing the bottom end. An outline schematic of an upending operation is shown in Fig. 2.22.

Figure 2.21 Gantries Used to Rotate a Suspended Load *(J&R Engineering Co., Inc.)*

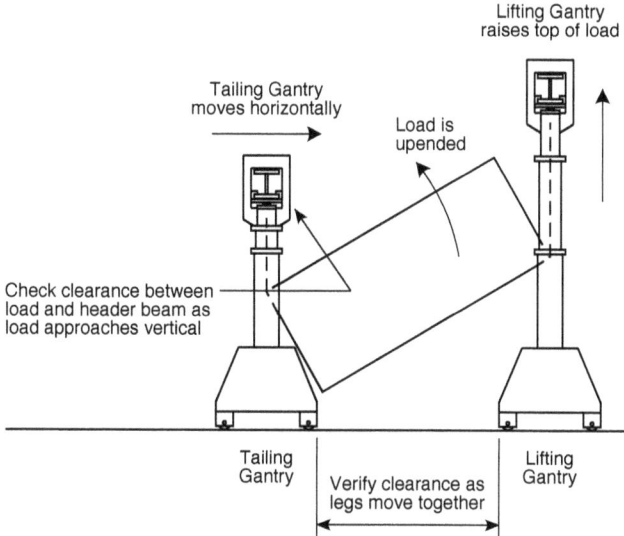

Figure 2.22 Upending a Load with Two Two-Leg Gantries

Noteworthy in Fig. 2.22 are the cautions regarding clearances between the gantry leg bases as the legs move closer together and between the load and the tailing gantry header beam as the load approaches vertical. Gantry legs are generally quite sensitive to horizontal loading. As such, planning on allowing the rigging to drift out of plumb in order to compensate for inadequate clearances should be avoided.

The upending or laying over of a load can also be performed with a single two-leg gantry system. In this operation, the top end of the load is raised or lowered by the gantry and the lower end of the item bears on the floor or ground (Fig. 2.23). This approach requires that the item being handled is capable of being supported in this manner and that the ground or floor is strong enough to safely support the concentrated load at the bottom end (i.e., the tailing reaction). The use of steel plates to provide a hard bearing/rolling surface under the lower end of the load is common practice. Additionally, the single two-leg gantry must be operated to simultaneously lift and travel as the load is rotated.

Upending or downending a load with hydraulic gantries brings with it some of the same considerations required in the planning of such a lift using mobile cranes. Three issues, in particular, are noted here.

First, the load distribution between the two two-leg gantries (or between a single two-leg gantry and the ground or floor) must be well understood. This distribution often changes as the lifted load is rotated, depending on the geometry of the lift points relative to the location of the center of gravity. Calculation of this load distribution is a required part of lift planning.

Second, positive control of the load as the center of gravity passes over the lower pivot point must be provided (Fig. 2.23). An abrupt movement of the

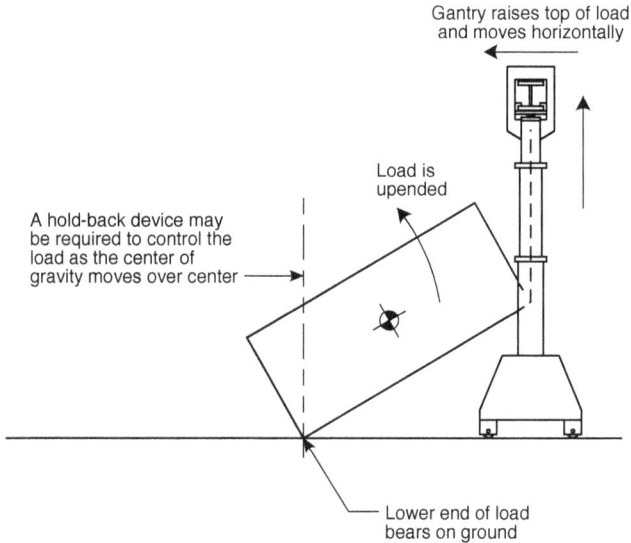

Figure 2.23 Upending a Load with One Two-Leg Gantry

load can occur during rotation if the load is not adequately controlled. Such a movement will impart significant horizontal forces into the lifting gantry, possibly causing a collapse (either toppling of the gantry system or a structural failure of one or more components of the system). Attempting to "catch" a rotating load while downending can also impart large vertical impact loads into the system and should never be a planned part of a lift. Continuous control of the load is an absolute imperative.

Third, the combined horizontal and vertical movements of the gantry system throughout the lift must be carefully coordinated to maintain the rigging as close to plumb as possible in order to minimize the development of horizontal forces (refer to Chapter 4 for a discussion of how out-of-plumb rigging results in the imparting of horizontal forces to the gantry system).

2.2.5   Load on Top of the Header or Cross Beams

The most common lift applications as described in the preceding sections have the lifted load suspended below the header or cross beams, typically using lift links and conventional rigging hardware, as has been illustrated. A gantry system may also be used to raise a load from below. That is, the load is carried on the tops of the header beams or cross beams (Fig. 2.24). All of the previously discussed movements, such as system travel and side shifting, can be performed in this configuration, although the hardware needed to side shift or rotate the load will most likely be of a design markedly different than that which has been discussed previously for use with a suspended load.

Figure 2.24 Lifting a Load on Top of the Cross Beams *(Kenneth A. Goodgame)*

Lifting in this manner also places unique demands on the lifted load. Without defined lift points, such as padeyes or trunnions, the ability of the lifted load to be supported on the header or cross beams must be verified. This procedure is most readily adaptable to loads such as skid-mounted machinery, vessels carried on saddles, or items that are uniformly rigid, such as large box girders or other heavy steel fabrications.

Gantry systems are most commonly used to lift the load from below when the available headroom is not adequate to allow lifting from above. The lift of a Yankee dryer seen in Fig. 2.24 used this technique for a different reason. Note that this lift used two gantry systems. The outer gantry system lifted the dryer high enough to allow positioning of the inner gantry system. The dryer was then set down on the cross beams of the inner gantry system. The rigging and header beams of the outer gantry system were removed to allow further lifting of the dryer using the inner gantry system.

The four cross beams of the inner gantry system are actually tracks for a skidding system. The inner gantry then continued the lift with the dryer supported on the tops of the cross beams. Once the cross beams were level with a mating skidding surface supported on the building floor at the wall opening, the dryer was skidded off the cross beams and into the building. This operation required that the cross beams be horizontally secured to avoid side loading the gantry legs when skidding the dryer.

2.3   ADVANCED GANTRY SYSTEM APPLICATIONS

Rigging contractors occasionally use hydraulic gantry systems as a part of a more complex lifting system. These applications often combine the hydraulic gantry system with other types of lifting equipment to provide a complete functionality that is not provided by either piece of equipment alone. The engineering, planning, and execution of a lift of this nature require an appropriate level of expertise both with the gantry system and with the mated equipment. Five such advanced applications are discussed here, two of which involve pairing the gantry system with another type of lifting equipment to provide examples of the versatility of the hydraulic gantry system when used as a part of a larger lifting system.

2.3.1   Strand Jacks Supported on a Gantry System

A growing number of specialty lifting and rigging contractors use hydraulic gantry systems in conjunction with strand jacks for heavy lifting (Fig. 2.25). This equipment combination provides two significant benefits. First, the strand jacks provide an almost unlimited lifting height, something that is far in excess of the capability of the gantry system. Second, the gantry system's propel system provides a travel capability without the need on the part of the contractor to design and build a drive mechanism. This horizontal movement advantage can be further increased by supporting the strand jack beams on side shifting devices, thus providing lateral, as well as longitudinal, movement using the gantry system functions. The overall capability of such a system is much like that of a high-capacity overhead crane.

The most common hydraulic gantry / strand jack configuration is that shown in Fig. 2.25. A fixed structural frame provides support for the gantry system track over its required length of travel and at the necessary elevation. The gantry system carries an arrangement of header and cross beams on which are mounted the strand jacks. The strand bundle lower anchors are then rigged to the load to be lifted, either directly or through some arrangement of lifting beams, as dictated by the handling requirements of the load.

Figure 2.25 Strand Jacks Supported by a Gantry System *(Burkhalter Rigging, Inc.)*

Additional efficiency can be developed in this hydraulic gantry / strand jack arrangement if the gantry's hydraulic system can also be used to power the strand jacks. Such an integration will further simplify the setup and operation of the lifting equipment.

2.3.2   Combination Lifts

The second type of lifting application to be highlighted here is the combination lift, an arrangement in which the hydraulic gantry system is used in conjunction with another type of lifting equipment, such as a crane or forklift. Lifts of this type are particularly complex since the hydraulic gantry system and the other piece of lifting equipment most likely behave in different ways. Managing these differences requires a detailed knowledge of both types of equipment to assure that the gantry system and the equipment with which it is paired will work together. Coordination of movement to minimize side loading to either piece of equipment is essential.

Figure 2.26 shows a four-leg hydraulic gantry system used as a tailing device for the upending of a tall process vessel. The lifting of the top of the vessel is performed by a tower-supported strand jack system. As with any upend / tail lift, the horizontal movement of the gantry system must be coordinated with the vertical movement of the top of the load.

2.3.3   Unconventional Component Arrangement

The gantry system shown in Fig. 2.27 utilizes conventional top side equipment (header beams, side shift trolleys, and cross beams) in an unusual manner. The two cross beams are mounted on top of the side shift trolleys, thus allowing the pair of

Figure 2.26 Gantry System Used as a Tailing Device *(David Duerr, P.E.)*

Figure 2.27 Cross Beams Mounted on Side Shift Trolleys *(PSC Crane & Rigging)*

beams to move laterally along the header beams, either as a unit or individually. This arrangement provides flexibility in the initial positioning of the four lift links, as well as the ability to side shift the suspended load.

2.3.4   Barge-Mounted Gantry System

The next advanced gantry system application to be discussed is the use of a gantry system on a floating barge (Fig. 2.28). In conventional practice, a fundamental rule of gantry system setup is that the supporting surface must be level and firm. The deck of a floating barge clearly does not meet this requirement. Wind, current, and movement of the lifted load will all cause the barge to roll (rotate about its longitudinal axis) or pitch (rotate about its transverse axis). These motions will

Figure 2.28 Gantry System on a Floating Barge *(Crane Rental Corporation)*

cause the barge deck to move out of level and will generate inertial forces that the gantry system and its supports must resist.

The use of a gantry system on a floating barge requires detailed engineering and planning that address the barge's response to the operation, as well as all of the normal lift planning issues. Because of the potentially significant horizontal loads that may act on the gantry system, the gantry manufacturer should be asked to evaluate the system for the planned lift (as was done for the project shown in Fig. 2.28).

A more complex gantry operation is illustrated in Fig. 2.29. Here, the gantry system, still being assembled in the photo, spans between the dock and a support barge. The lift will move a piece of cargo from the dock to the cargo barge that is moored along the bulkhead. The two waterside gantry legs are mounted on the support barge that is moored outboard of the cargo barge. The landside gantry legs will be firmly supported in the conventional manner and will be relatively unyielding during the lift. The waterside legs will carry a continually increasing share of the load as the cargo item is side shifted from dock to barge, thus changing the draft of the support barge. Placement of the gantry legs on the support barge, along with the use of ballast, should be designed minimize changes in list and trim. Operation of the system during the lift must be controlled to keep the header beams as level as possible.

The planning and engineering of either of these types of lifts require a working knowledge of basic naval architecture principles to facilitate calculation of the support barge movements in response to the changing loads imposed by the gantry system and to design a mooring system that will adequately maintain the support barge's position.

A less complex barge application is the use of a gantry system on a grounded barge. In this case, the barge will not move under the effects of the lifting operation. The planning and engineering must consider only the structural capacity of the deck and internal structure of the barge and the ballasting of the barge to assure

Figure 2.29 Gantry System Spanning from Dock to Barge *(Norris Brothers Co., Inc.)*

that it will remain firmly grounded under all conditions of loading and extremes of tide.

2.3.5   Lifting with More Than Four Gantry Legs

Lifting a load with two or four gantry legs is relatively routine. These lifts are common enough that experienced gantry system users are generally comfortable planning and performing such lifts. This may change when lifting with a greater number of legs.

The load pictured in Fig. 2.30 is being lifted from eight trunnions using eight gantry legs. The complexity of a lift like this demands an increased level of engineering and planning to assure that the load path is understood, that all of the gantry system components can safely carry the imposed loads, and that the gantry operator knows what to expect in terms of how much load will be carried by each gantry leg. Complexity in performing the lift may also be amplified by limitations in the ability of the control system to coordinate the functions of more than four legs. This places an even greater burden on the gantry operator.

In recent years, some of the gantry system manufacturers have introduced advanced systems capable of simultaneously controlling more than four gantry legs for both lift and travel. Some systems can even control different models of gantry legs working together in one arrangement. The use of control system with this level of capability when lifting with more than four legs provides an obvious benefit to the contractor.

The lift arrangements and applications described in this chapter are simply the more common uses of telescopic hydraulic gantries for heavy and specialized lifting. This is not intended to be an all-inclusive list of the ways that gantries can (or should) be used, but rather just a snapshot of the big picture of gantry use.

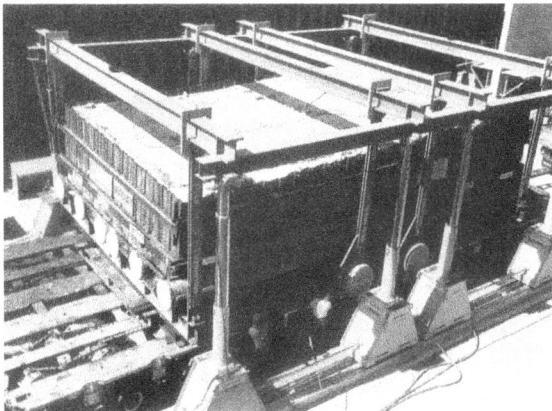

Figure 2.30 Lifting a Load with Eight Gantry Legs *(Lift Systems, Inc.)*

2.4   CONTACT INFORMATION

Throughout this book, a common thread among many of the discussions calls for the gantry system user to follow information or abide by recommendations provided by the manufacturer. This is necessary due to the relatively low level of standardization in the design and performance of hydraulic gantry systems at present.

The operator's manual is, of course, the first place to look for information with respect to a particular gantry system. Additional information is often available through the manufacturers' web sites. Following are the web addresses and locations for the five major manufacturers of telescopic hydraulic gantry systems (current as of this writing).

Enerpac Integrated Solutions B.V. (formerly Hydrospex Cylap B.V.)
Hengelo, The Netherlands
www.enerpac.com/en/integrated-solutions/hydraulic-gantries

Greiner GmbH–Fahrzeugtechnik
Neuenstein, Germany
www.greiner-fahrzeugtechnik.de/hubsysteme_en.html

J&R Engineering Co., Inc.
Mukwonago, Wisconsin, United States
www.jrengco.com

Lift Systems, Inc.
East Moline, Illinois, United States
www.lift-systems.com

Riggers Manufacturing Company
Dixon, Illinois, United States
www.riggers.com

3 The Hydraulic System

The heart of a telescopic hydraulic gantry is its hydraulic system. The details of the hydraulic system vary from one gantry model to the next, depending on the number of cylinders in each gantry leg, the use of hydraulically actuated boom locking devices, the type of propel mechanism used, and the type of control system. However, the basic components and circuitry are common among the different models. This chapter provides a general discussion of the functions of the primary hydraulic components and the operation of the basic gantry hydraulic system.

This chapter is not meant to be an all-inclusive text on hydraulic systems and components. Rather, the intent is to provide a general discussion of the subject as applicable to hydraulic gantry systems, thus providing to the reader a basic understanding of the equipment and the principles of operation.

3.1 SYSTEM COMPONENTS

The functions of each of the major hydraulic components are described in this section. With this background established, the operation of the complete hydraulic system is then examined in Sections 3.2 and 3.3.

3.1.1 Telescopic Cylinders

The earliest hydraulic gantry systems built in the 1960s and 1970s used single stage cylinders (refer to Chapter 1 for historical information). The gantry system shown in Fig. 1.5, which was built in the late 1970s, was the first to use multiple stage telescopic cylinders, thus opening the door to the extended-to-retracted height ratios that make gantries the versatile tools that they are today. From that point forward, most hydraulic gantries have been built with telescopic cylinders of two or more stages. Further, the cylinders used in most gantries are double acting. That is, fluid pressure is required below the pistons to extend the cylinder and above the pistons to retract the cylinder.

The main components of a basic two-stage telescopic hydraulic cylinder are illustrated in Fig. 3.1. As with typical single stage cylinders, the outermost section is called the barrel and the innermost is called the rod. The intermediate sections

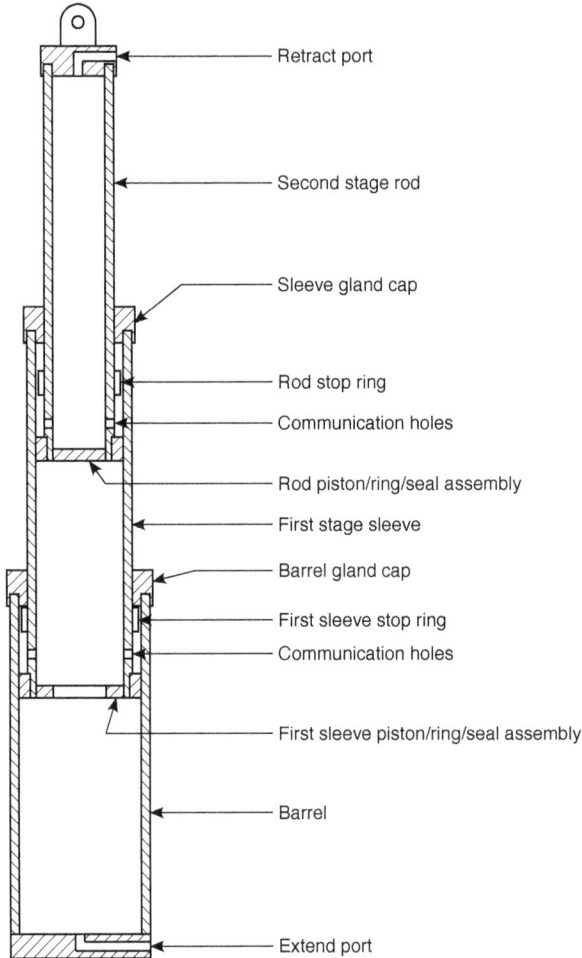

Figure 3.1 Telescopic Cylinder

are called sleeves. Of particular importance to the structural integrity of the cylinder are the stop rings on the sleeves and rod. The stop rings limit the travel of each sleeve or the rod, thus maintaining a minimum length of overlap between the sections of the cylinder. A key feature with respect to the hydraulic functioning of the cylinder is the set of communication holes near the bottom of each sleeve and the rod. These holes allow fluid to move into and out of the annulus (also called the annular space) between adjacent sections as the cylinder extends and retracts.

The mechanical connection of the cylinder to the gantry leg varies, depending on the leg style. The cylinder for a bare cylinder gantry leg will typically have a heavy flange plate welded to the barrel (a feature not shown in Fig. 3.1). This flange plate bears on the top of the leg's base weldment and provides both a vertical and a rotational connection between the cylinder and the base. The header plate is pinned to a lug or clevis at the end of the rod. The cylinder within a gantry leg boom is often mounted with the rod down to facilitate connection of the cylinder sections to the boom sections. The rod will be connected at the bottom of the boom using a pinned joint or other articulating attachment. If the boom is fully powered, the barrel will be connected to the uppermost boom section with a similar bolted or pinned connection. If the leg has a manual section, the barrel connection typically uses pins or bars that the gantry user can manipulate to allow extension of the manual boom section, retraction of the cylinder after the manual section has been locked to the adjacent boom section, and then connection of the cylinder to the next lower boom section for further extension of the leg. Details of these devices are best found in the literature and operator's manuals of the specific gantry models.

The flow of hydraulic fluid through a telescopic cylinder is illustrated in Fig. 3.2 for cylinder extension and in Fig. 3.3 for cylinder retraction. During cylinder extension, fluid enters through the extend port and moves the sleeve and rod together (Fig. 3.2*a*). When the sleeve reaches the limit of its stroke and its stop ring contacts the inside surface of the barrel gland cap (Fig. 3.2*b*), the rod begins to extend relative to the sleeve (Fig. 3.2*c*). Fluid in the annular spaces above the sleeve and rod pistons flows out through the retract port.

During retraction, fluid enters the cylinder through the retract port and passes through the communication holes at the base of the rod into the annulus between

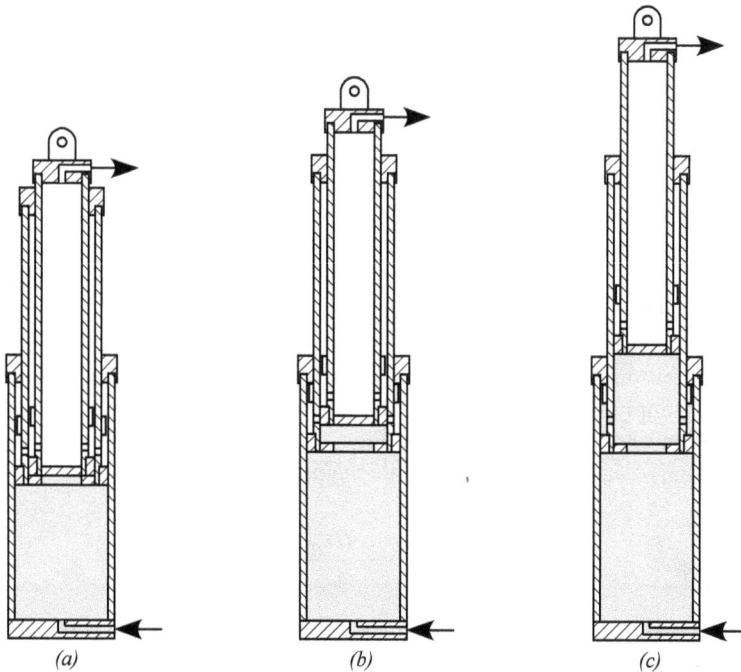

(a)   *(b)*   *(c)*

Figure 3.2 Hydraulic Fluid Movement During Cylinder Extension

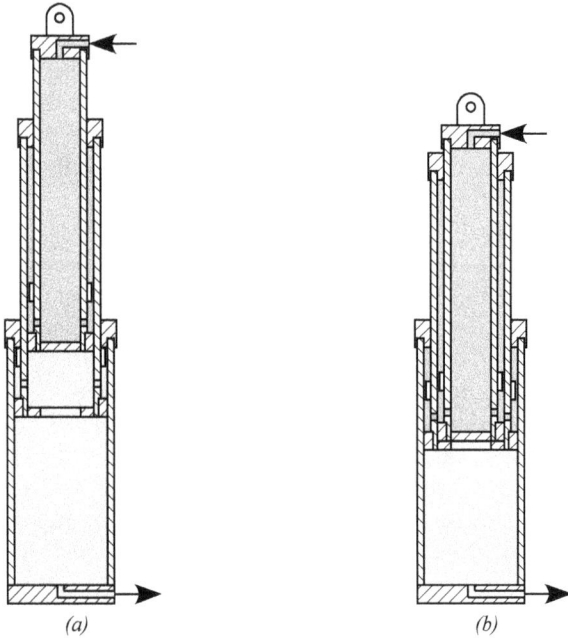

Figure 3.3 Hydraulic Fluid Movement During Cylinder Retraction

the rod and the sleeve (Fig. 3.3a). The fluid below the pistons flows out through the extend port as the cylinder retracts. The fluid pressure acting in the annulus causes the rod to retract. Once the rod has fully retracted, the sleeve begins to retract (Fig. 3.3b).

Understanding the operation of the hydraulic system and the relationship between the fluid pressure and the supported load requires knowing the fluid pressures above and below each piston, both during cylinder extension and retraction. A method by which the cylinder flow pressures can be calculated is discussed in Section 3.3. Here, we will identify the cross-sectional areas within the cylinder that will be needed for those calculations.

The areas used in the cylinder fluid pressure calculations, as illustrated in Fig. 3.4, are the piston area A_B and the annular area A_A. The piston area, as the name implies, is simply the cross-sectional area below the piston of the section under consideration. For the detail shown in Fig. 3.4 and letting ID_B equal the barrel's bore (internal) diameter, we can calculate the piston area using Eq. 3.1.

$$A_{B1} = \frac{\pi \, ID_B^2}{4} \tag{3.1}$$

The annular area is the net area above the piston of the section under consideration. Letting OD_S equal the outside diameter of the first sleeve, we can calculate the annular area between the barrel and the first sleeve using Eq. 3.2.

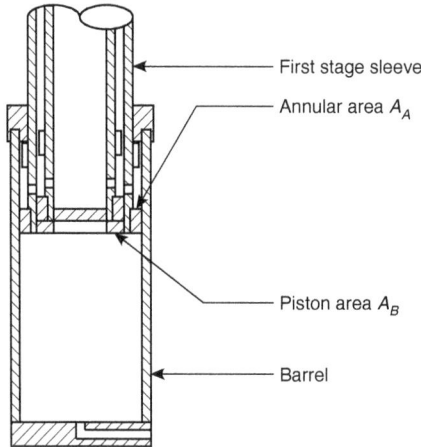

Figure 3.4 Details of Cylinder Cross-Sectional Areas

$$A_{A1} = \frac{\pi \left(ID_B^2 - OD_S^2 \right)}{4} \tag{3.2}$$

The "1" added to the subscripts of A_B and A_A in Eqs. 3.1 and 3.2 indicates that the values are applicable to the piston and annular areas of the first sleeve. Similar changes in the subscripts can be made to denote applicability to additional sleeves and the rod.

As discussed on page 49, telescopic boom gantry legs are often configured with the cylinder installed with the rod end at the bottom. In this arrangement, it is desirable to locate both the extend and retract ports at the end of the rod. The required fluid flow path, as illustrated in Figs. 3.2 and 3.3, from the extend port is maintained by installing a tube in the rod that runs from the extend port to the rod piston, thus connecting the extend port with the volume below the rod piston. The areas calculated with Eqs. 3.1 and 3.2 are unchanged, regardless of such differences in the internal piping of the cylinder.

3.1.2   Check Valves

The simplest of the directional control valves we will discuss is the check valve. Fig. 3.5a is a graphical illustration of a poppet-type check valve. Fig. 3.5b is the symbol used in a hydraulic circuit diagram to represent a basic check valve that either uses a very light spring to hold the poppet in place or that has no spring at all. Fig. 3.5c represents a check valve that has a stronger spring sized to resist flow up to a defined pressure. The check valve serves the simple function of restricting fluid flow to one direction. The pressure of fluid entering the valve through Port 1 will displace the poppet, allowing flow out through Port 2. Flow in the reverse

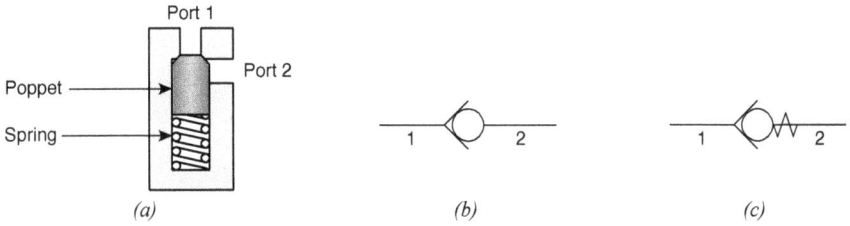

Figure 3.5 Check Valve

direction (in through Port 2 and out through Port 1) is not possible. As soon as pressure at Port 1 drops off, the spring will push the poppet back toward Port 1, thus blocking the reverse flow.

Styles of check valves other than that shown use either a round ball or a flapper as the flow control element. Further, the check valve may or may not have a spring, as shown in Fig. 3.5a, to hold the flow control element in position. Check valves without a spring depend on the fluid pressure or gravity (when mounted vertically with Port 1 down) to maintain the position of the element. Regardless of the specifics of the hardware, the basic function of the check valve and its use in a hydraulic circuit remain the same.

[The illustrations of the various hydraulic system components provided in this section are drawn to show clearly the functions of the components. These figures are not necessarily representative of actual fluid power products. Specific product information is best obtained from the component manufacturers.]

3.1.3   Relief Valves

The relief valve (Fig. 3.6) is a normally closed pressure control valve that serves to limit the maximum fluid pressure in a hydraulic system. The system pressure is applied to the relief valve through Port 1. The poppet is held in the closed position by the spring. The compression load in this spring is set by means of the adjusting

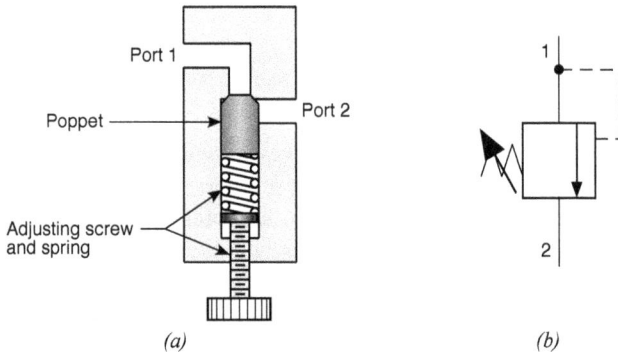

Figure 3.6 Relief Valve

screw in order to establish the maximum fluid pressure at Port 1 that the valve can resist. If the line pressure exceeds the pressure setting of the relief valve, a value often referred to as the cracking pressure, the poppet is displaced, allowing fluid to pass through the valve body and out through Port 2. A line from Port 2 is typically routed back to the hydraulic fluid tank to collect the fluid that passes through the valve.

The relief valve pressure is normally set to a value somewhat greater than the line's design operating pressure. Such a setting provides protection against excessive pressure without compromising the system's operation.

3.1.4   Counterbalance Valves

The counterbalance valve (Fig. 3.7) is a pressure control valve that allows the free flow of fluid in one direction and blocks flow in the opposite direction until a certain preset pressure is reached. As used in a gantry system, the counterbalance valve prevents the lift cylinder from retracting at an uncontrolled rate due to the weight of the supported load. Because of this function, the counterbalance valve is frequently referred to as a holding valve or a lock valve. The counterbalance valve only permits flow from Port 1 to Port 2 when the force on the spool developed by the pressure of the oil at Port 1 and through pilot Port 3 exceeds a defined level that is set by means of the adjusting screw and spring. This pressure level, again called the cracking pressure, is set by the gantry system manufacturer to a value that is somewhat greater than the pressure created by the maximum load that is supported by the cylinder due to the rated load of the gantry leg, much like the relief valve setting previously discussed.

The counterbalance valve shown in Fig. 3.7 is referred to as a pilot-assisted valve due to its use of both an internal pilot and an external pilot (Port 3) to control

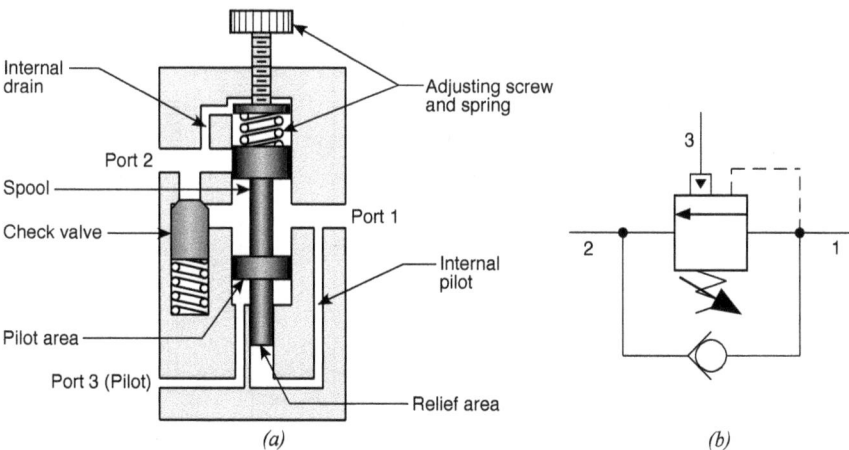

Figure 3.7 Counterbalance Valve

the movement of the spool. This combination of internal piloting and external pilot assist provides for smooth operation of the cylinder when lowering the load and a more efficient use of system fluid pressure, as compared to just internal or external piloting alone.

For maximum reliability, the counterbalance valve should be installed directly on the cylinder close to the extend port to eliminate any fittings or flexible lines between the valve and the cylinder. (The failure of such parts would cause a loss of pressure in the cylinder and a resulting uncontrolled retraction of the cylinder.) The most commonly used style of counterbalance valve in hydraulic gantry applications is the cartridge type, where the valve mechanism screws into a machined steel body. This body can be welded directly to the cylinder.

The function of the counterbalance valve can be best seen by following the fluid flow and by observing the spool and check valve positions (Fig. 3.8). To extend the cylinder (Fig. 3.8a), hydraulic fluid enters the counterbalance valve through Port 2, passes through the check valve, exits the counterbalance valve through Port 1, and enters the cylinder barrel. The spool of the counterbalance valve is fully seated when the fluid flows in this direction.

Fig. 3.8b shows the spool and check valve positions when holding the load. Fluid pressure from the cylinder acts on the counterbalance valve through Port 1. This pressure acts on the piston below the spool through the internal pilot, but is fully resisted by the spring. In this position, the counterbalance valve functions much like the relief valve described above, where the spring is adjusted to enable the counterbalance valve spool to remain fully seated at the design operating pressure that acts at Port 1.

To retract the cylinder, fluid must be able to flow into the counterbalance valve through Port 1 and out through Port 2. This requires that the spool be shifted as shown in Fig. 3.8c. The spool is shifted by a combination of fluid pressure applied to Port 3 (the external pilot) and an increase in the cylinder pressure created

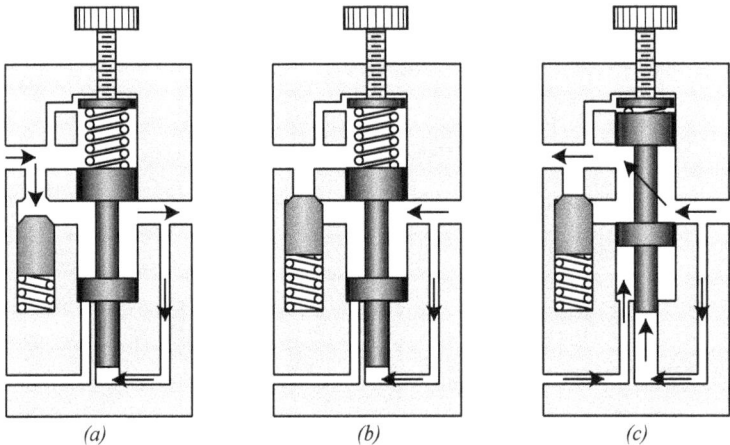

(a)   (b)   (c)

Figure 3.8 Counterbalance Valve Positions

by introducing a fluid pressure into the retract port of the cylinder. That is, the cylinder does not simply retract due to the weight of the supported load. Rather, it is pushed down by the application of fluid pressure. Because the opening of the counterbalance valve for cylinder retraction is controlled by fluid pressures that are, in turn, controlled by the gantry operator, we can see that the use of the counterbalance valve provides the operator with the ability to smoothly and positively control the retraction of the loaded cylinder.

The shifting of the spool as described depends on a combination of internal (Port 1) and external (Port 3) pilot pressures and the pressure setting of the adjusting spring. The area on which the external pilot pressure acts is called the pilot area. The area on which the internal pilot pressure acts is called the relief area. The ratio of pilot area to relief area is called the pilot ratio. Valve specifications typically give the pilot ratio, rather than the values of the two areas.

The fluid pressures acting at the counterbalance valve ports, the valve setting, the pilot ratio, and the cylinder areas are related as follows for a system with the cylinder oriented vertically and supporting a load with the cylinder in compression. First, we can identify the following variables.

P	=	pilot pressure at Port 3 required to open the valve;
S	=	pressure setting of the counterbalance valve (cracking pressure);
L	=	pressure to Port 1 due to the supported load;
A_P	=	pilot area;
A_R	=	relief area;
P_R	=	valve pilot ratio = A_P / A_R;
ID_B	=	cylinder bore (internal) diameter;
A_B	=	cylinder piston area;
OD_S	=	cylinder sleeve (or rod) outside diameter;
A_A	=	cylinder annular area; and,
C_R	=	cylinder area ratio = A_B / A_A.

Note that the valve pressure setting S is the pressure at Port 1 that will shift the spool in the absence of any pilot pressure at Port 3. That is, the pressure setting S is the valve's holding pressure. The force required to open the counterbalance valve in the absence of pressure at Port 3 is simply SA_R.

When the cylinder is hydraulically retracted, a pressure P is applied to the counterbalance valve at Port 3. Given the hydraulic circuitry common to gantry legs, this same pressure is applied at the cylinder's retract port, thus applying a pressure above the cylinder piston. This fluid pressure applied above the cylinder piston increases the pressure below the piston and, correspondingly, at Port 1 of the counterbalance valve. The pressure below the piston due to the applied pressure above the piston is termed L_P and is calculated using Eq. 3.3.

$$L_P = \frac{PA_A}{A_B} = \frac{P}{C_R} \tag{3.3}$$

The total pressure below the cylinder piston is equal to the pressure due to the supported load plus the pressure calculated using Eq. 3.3. That is, the total pressure L_T is equal to $L + L_P$.

With these variables and relationships defined, we can now write Eq. 3.4 from which the pilot pressure at Port 3 required for cylinder retraction can be calculated.

$$P = \frac{S - L}{P_R + 1/C_R} \tag{3.4}$$

The use and validity of these equations can best be illustrated by means of an example problem.

EXAMPLE 3-1

Consider a cylinder with a bore diameter of 12.000 inches and a rod outside diameter of 10.500 inches. A load of 225,000 pounds is supported. The counterbalance valve has a pilot area $A_P = 0.60$ square inch and a relief area $A_R = 0.20$ square inch, thus giving a pilot ratio P_R of 0.60 / 0.30 = 3.00. The counterbalance valve has a setting S of 2,500 psi.

The piston area A_B (based on the bore diameter) is 113.10 square inches, so the pressure L at Port 1 equals 225,000 / 113.10 = 1,989 psi. The annular area above the piston A_A is 26.51 square inches, so the cylinder area ratio C_R is equal to 113.10 / 26.51 = 4.27. We can now apply Eq. 3.4 to determine the pilot pressure P required to retract the cylinder.

$$P = \frac{2,500 - 1,989}{3.00 + 1/4.27} = 158 \text{ psi}$$

To verify this value, we can calculate the various pressures and forces acting within the valve, as follows.

With no pilot pressure acting at Port 3, a pressure of 2,500 psi acting at Port 1 is required to crack the valve. Thus, the force required to compress the adjusting spring is 2,500 x A_R = 2,500 x 0.20 = 500 pounds. Eq. 3.3 is used to calculate the pressure below the piston that is developed by the pressure P = 158 psi acting above the piston.

$$L_P = \frac{158}{4.27} = 37 \text{ psi}$$

Thus, the total pressure below the piston and, therefore, at Port 1 of the counterbalance valve, is 1,989 + 37 = 2,026 psi and the force to the spring due to this pressure acting on the relief area is 2,026 x 0.20 = 405 pounds. The pressure P = 158 psi acting on the pilot area A_P = 0.60 square inch creates a force to the spring of 158 x 0.60 = 95 pounds for a total force to the spring of 405 + 95 = 500 pounds, identical to that calculated above.

This example can be used to illustrate another value of the external pilot. $L = 0$ psi with no load on the cylinder. Solving Eq. 3.4 for this case gives us a pilot pressure of 773 psi to retract the unloaded cylinder. Without the external pilot, $P_R = 0$, so a pressure to the top end of the cylinder of 10,667 psi would be required to develop a pressure of 2,500 psi below the cylinder piston (and at Port 1) to overcome the counterbalance valve setting and retract the cylinder. This pressure is obviously excessive relative to the operating pressure required to lift a load.

Hydraulic gantries are often used to support a lifted load for hours or even days. In this circumstance, the fluid pressure L must be held in the cylinder by the counterbalance valve. All spool-type valves exhibit some leakage, so if the load is to be held on the cylinders for an extended period of time, this leakage must be considered.

Valve leakage is commonly defined in terms of drops of oil per minute. One valve manufacturer defines a drop of oil as the volume for which 250 drops equals one cubic inch, or 15 drops equals one cubic centimeter. Leakage through a valve varies from one model to the next and from one manufacturer to the next. Specifications for a specific valve must be obtained from the valve manufacturer for investigation of a particular installation.

The significance of leakage through the counterbalance valve can be illustrated by example using the cylinder described above.

EXAMPLE 3-2

Consider a cylinder with a bore diameter of 12.000 inches. The counterbalance valve has a leakage rate of 5 drops per minute, as specified by the valve manufacturer. Calculate the length of retraction under load in 24 hours.

5 drops per minute = 5 x 60 x 24 = 7,200 drops in 24 hours

7,200 drops in 24 hours / 250 drops per cubic inch = 28.80 cubic inches in 24 hours

The cylinder piston area $A_B = \pi\ 12.000^2 / 4 = 113.10$ square inches

Cylinder retraction rate = 28.80 / 113.10 = 0.25 inch in 24 hours

The rate of leakage of oil through a valve as reported by the valve manufacturer is typically measured in a system with extremely clean oil. A valve in service can become fouled by contamination in the oil, which can increase the leakage rate. This point highlights the importance of keeping the system clean and of recognizing the effects of in-service conditions on the performance of valves and other hydraulic components.

The retraction of a hydraulic cylinder under load due to fluid leakage is referred to as *drift* in much of the literature that is specific to industrial hydraulic systems and components. With respect to hydraulic gantries, the term drift refers to the transverse movement of a gantry leg's telescopic lift boom that occurs due to clearances between the boom sections. This is discussed in detail in Chapter 5. Due to this gantry-specific usage of the term, cylinder retraction due to fluid leakage is not referred to as drift here.

3.1.5   Shuttle Valves

The shuttle valve (Fig. 3.9) is a directional control valve that is used to select the higher of two inlet fluid pressures and connect that inlet port with the outlet port

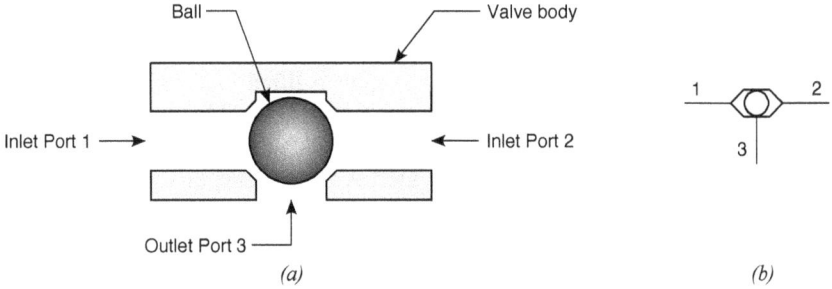

Figure 3.9 Shuttle Valve

of the valve. As illustrated in Fig. 3.9, the ball will shift to the left or to the right within the valve body, depending on which inlet port pressure is greater. The fluid from the inlet port with the higher pressure will flow through the valve and out Port 3 and flow from the inlet port with the lower pressure will be blocked. We can see that if the relationship of the two inlet pressures changes during operation, the shuttle valve ball position also changes, thus keeping the higher pressure flow connected to the outlet port at all times.

3.1.6   Flow Restrictors

A flow restrictor, in its most basic form, is simply a fitting with an orifice of a diameter that is smaller than that of the line in which it is installed. The reduction in the cross sectional area of the flow passage serves to restrict the rate of fluid flow to some reduced level. The flow rate through the restrictor is a function of the diameter of the orifice, the fluid's viscosity, and the pressure.

A second type of flow restrictor is adjustable, using an adjustment screw to alter the area of the orifice, thus allowing the system operator to change the maximum flow rate.

Fig. 3.10a is the circuit diagram symbol for a fixed flow restrictor. Fig. 3.10b is the symbol for an adjustable restrictor.

3.1.7   Directional Control Valves

The operator controls the fluid flow through the system by means of one or more directional control valves. A basic four-way directional control valve is illustrated

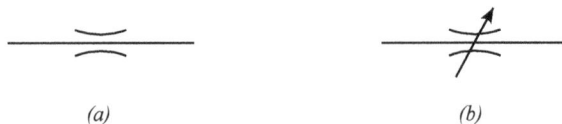

Figure 3.10 Symbols for Flow Restrictors

in Fig. 3.11. The identifier "four-way" indicates that the valve controls fluid flow among four ports. The symbol shown in Fig. 3.11*b* is for a valve that is manually controlled by the operator. The spool of a directional control valve may also be shifted by means of an electrical solenoid or by fluid pressure applied through a pilot port. The symbols for these two types of directional control valves are shown in Figs. 3.12*a* and 3.12*b*, respectively.

Fluid flow is controlled by the position of the spool. The neutral position is shown in Fig. 3.11*a*. In this position, the spool blocks off Ports A and B, thus blocking flow to or from the connected lines. Some valve designs also block flow through Ports P and T, while others permit circulation through the valve from the pump back to the tank. The two operating positions of the valve are shown in Fig. 3.13. With the spool shifted to the left (Fig. 3.13*a*), flow from the pump (Port P) is directed to the line connected to Port A and flow from the line connected to Port B is directed back to the tank through Port T. With the spool shifted to the right (Fig. 3.13*b*), flow from the pump (Port P) is directed to the line connected to Port B and flow from the line connected to Port A is directed back to the tank through Port T. The flow through the valve in either direction can be throttled by incrementally opening and closing the valve.

The simple directional control valve shown in Fig. 3.11 illustrates the flow control capability required for a gantry system and the symbol shown in Fig. 3.11*b* will be used later in the descriptions of the hydraulic system. However, this type of valve is not the best choice for use in a gantry system with a manually operated hydraulic system for one simple reason. For a given position of the spool, the rate of flow through the valve is proportional to the fluid pressure. Thus, if two gantry legs are controlled by two valves and the two legs are not each supporting the same load, holding the two valves at the same position will result in different rates of extension or retraction of the two legs. This behavior would make the gantry system very difficult to safely operate manually. However, a basic directional

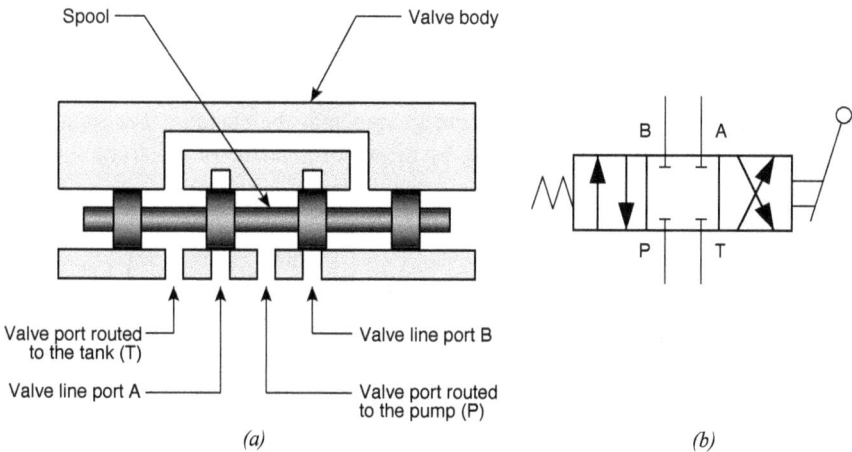

(a)   *(b)*

Figure 3.11 Directional Control Valve

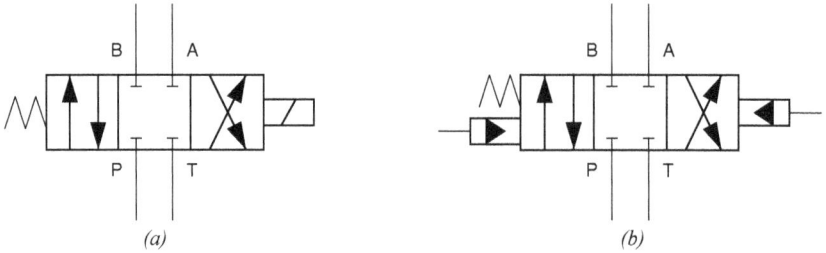

Figure 3.12 Alternate Directional Control Valve Symbols

control valve that is solenoid-actuated may be appropriate in an electric-over-hydraulic system in which the control system provides continual feedback to the valve as the cylinder extends and retracts.

Fig. 3.14 illustrates in symbol form a pressure-compensated directional control valve. This device utilizes a number of the previously described types of valves to maintain a constant flow through the valve regardless of the fluid pressure. This valve has the same four ports (P, T, A, and B) as the basic four-way valve and, from the gantry operator's point of view, functions in the same manner, with just a single actuator lever to control. However, unlike the basic valve, the gantry leg will always extend or retract at essentially the same rate for a given position of the valve's lever. This allows the operator to control and coordinate the multiple legs of a gantry system with greater reliability.

The pressure-compensated directional control valve diagrammed in Fig. 3.14 is typical of this type of valve, but it is not the only product configuration that is available. Regardless of the physical make-up of the valve, however, the basic function is as described.

3.1.8 Pressure Gauges and Transducers

Pressure gauges are used throughout the hydraulic system to provide the operator with necessary information, such as pump output pressure, line pressure to each cylinder, and the like. These pressure gauges may be standard Bourdon tube analog gauges that provide a reading by means of a needle or electronic gauges

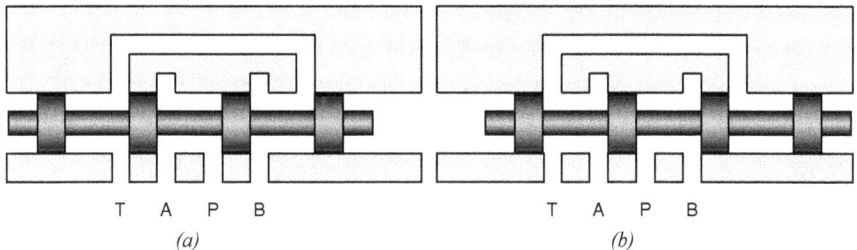

Figure 3.13 Directional Control Valve Spool Positions

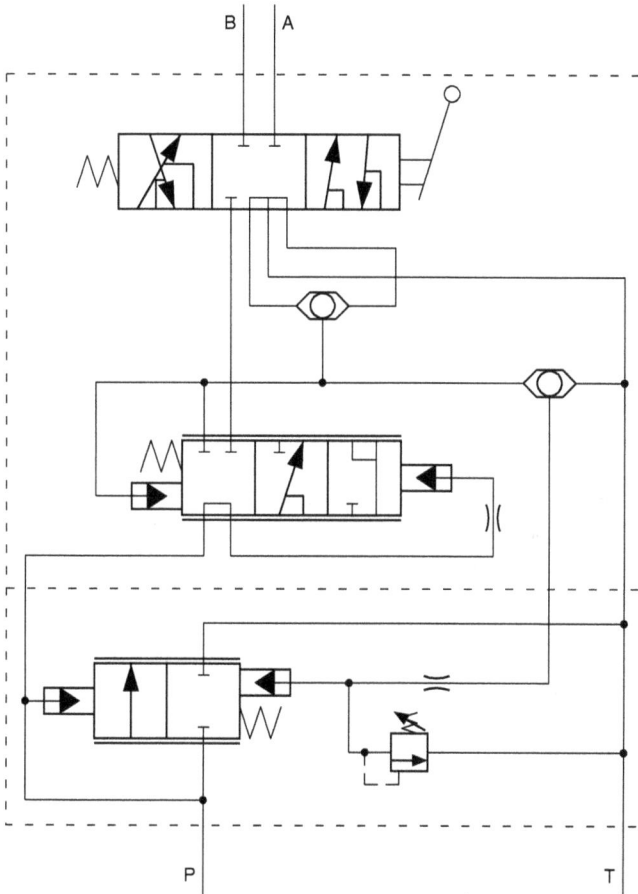

Figure 3.14 Pressure-Compensated Directional Control Valve

that use a transducer to measure the fluid pressure and provide a reading by means of a digital display on the operator's control unit.

The most significant use of pressure gauges in a gantry system occurs when the fluid pressure is used to determine the load being lifted by each gantry leg. When pressure gauges are used for this function, some industry guides (IHSA 2007, SC&RA 1996) recommend that the gauges or transducers for digital displays be mounted directly on the cylinders in order to eliminate deviations in the pressure readings due to losses through valves, fittings, and tubing. Of particular importance in such an installation is protecting the gauge or transducer from damage that could either impair its function or result in an uncontrolled loss of pressure, which could result in the dropping of a load.

Equating the fluid pressure in the cylinder to the load being lifted is not as simple as it sounds. The fluid pressure and load are directly related only while the load is being held in a static position. The need to force the fluid above the piston out of the cylinder will develop a back pressure due to flow resistance while the cylinder is being extended and the pressure required to open the cylinder's counterbalance valve while the cylinder is being retracted will similarly create deviations from this basic relationship. The methods by which these pressures can be calculated are discussed in Section 3.3.

3.1.9   Single Stage Cylinders

Single stage double acting cylinders are used in some gantry systems as part of the propel system (see Section 2.1.3 and Fig. 2.15) or as part of a side shifting mechanism (see Section 2.2.3). The cylinders used for these functions are of a common industrial design (Fig. 3.15) and, in fact, are often "off the shelf" products, rather than components that are specially designed for use as a part of a gantry system. Because these cylinders are standard products, a gantry system user may purchase cylinders for use in propel or side shift devices to meet the needs of a special lift. When doing so, a few points must be observed to assure getting the right component for the application.

The use of these cylinders for propel and side shift devices is very straight-forward, generally based simply on the manufacturer's load ratings and the stroke required for the intended application. Practical usage in either function dictates that the cylinder must be capable of supplying the required force in either direction, which tells us that selection is most commonly limited by the "retract" load rating. Keep in mind, however, that the retract load rating is typically based only on the operating pressure and the cylinder's annular area A_A. The "extend" load rating of a cylinder is the lesser of the operating pressure times the piston area A_B or the column buckling strength of the cylinder (NFPA 2010) divided by a suitable design factor. Buckling strength often limits the "extend" load rating of a long, slender cylinder.

Cylinders used for propel or side shifting ordinarily do not require a counterbalance valve; the extend and retract ports can be directly plumbed to the applicable directional control valve. However, if the cylinder is used in an application where it must be capable of holding a load, then use of a counterbalance

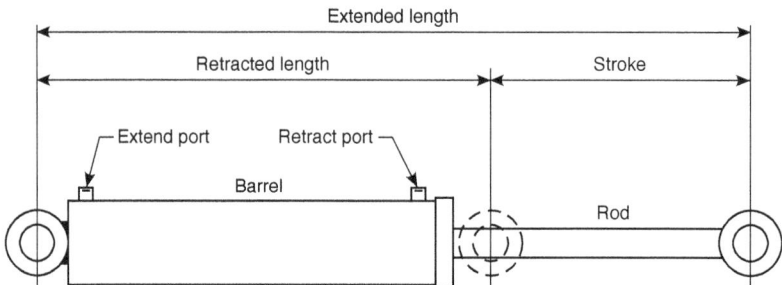

Figure 3.15 Single-Stage Cylinder

valve may be necessary. As always, each system design and component application must be evaluated based on its own demands.

3.1.10 Hydraulic Motors

The propel systems of many gantry legs use hydraulic motors (also called rotary actuators in some hydraulics textbooks), rather than push-pull cylinders. Both integral and external drive wheel propel systems use motors and either a gear train or a chain drive mechanism to transmit power from the motor output shaft to the drive wheels. Some side shifting lift links are also built with internal drives that use hydraulic motors, thus eliminating the need to install push-pull cylinders for each setup of the system.

Hydraulic motors are relatively compact and typically have just two ports for fluid connections (Fig. 3.16a). The motor symbol shown in Fig. 3.16b denotes a bi-rotation variable displacement motor. That is, the direction of rotation of the output shaft can be reversed by reversing the fluid flow through the motor and the displacement volume of the motor be adjusted by the operator to alter the speed of rotation. The bi-rotation characteristic is indicated by the two triangles within the circle; one triangle indicates a single-direction motor. The variable displacement characteristic is indicated by the diagonal arrow through the circle. This arrow is removed to indicate a fixed displacement motor. The speed of a fixed displacement motor is controlled by altering the fluid flow rate.

The hydraulic motors used in a gantry system are most commonly standard manufactured products that are selected for the particular application on the basis of the product ratings for output speed, horsepower, and torque. The gear trains or chain drives are then specified to develop the needed torque and slow speed at the drive wheels. These components are part of the gantry legs or side shift devices and rarely have to be selected or evaluated by the gantry system user. They are included in this section primarily for background information.

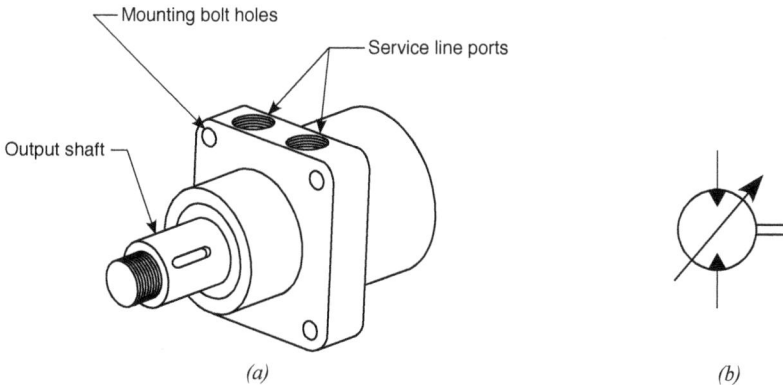

Mounting bolt holes

Service line ports

Output shaft

(a)

(b)

Figure 3.16 Hydraulic Drive Motor

3.2   THE HYDRAULIC CIRCUIT

Fig. 3.17 illustrates the basic hydraulic circuit for a single cylinder gantry leg and its control and power unit. Note that the complete hydraulic system of most gantries is more complex than that shown due to the need to include controls for the propel system, to provide pressure gauges, filters and other fittings, and on some models to provide hydraulic fluid flow control to multiple cylinders on a single leg. Telescopic boom gantry legs may also have hydraulically actuated boom locking devices. Further, the hydraulic circuit diagrams found in a gantry system operator's manual will typically be divided into separate diagrams for the gantry leg and for the control and power unit (for those systems with an external power unit).

The functions of the main hydraulic system components shown in Fig. 3.17 are as follows. The hydraulic pump draws oil from the tank and provides the operating fluid power to the gantry system. The pump and its internal combustion engine or electric motor are sized for the required oil flow rate and operating pressure. The directional control valve provides the gantry operator with the means to direct the flow of oil to the cylinder for extension or retraction. The counterbalance valves control the oil flow both when extending or retracting the cylinder and when holding a load. Last, the relief valves protect the system against excess fluid pressure.

The flow of the hydraulic fluid through this system is illustrated in Fig. 3.18a for extending the lift cylinder and in Fig. 3.18b for retracting the cylinder. The system component functions for extension and retraction are described in the following paragraphs.

To extend the cylinder, the directional control valve is shifted to send fluid from the pump (Port P) to the line at Port A. Fluid entering the valve at Port B

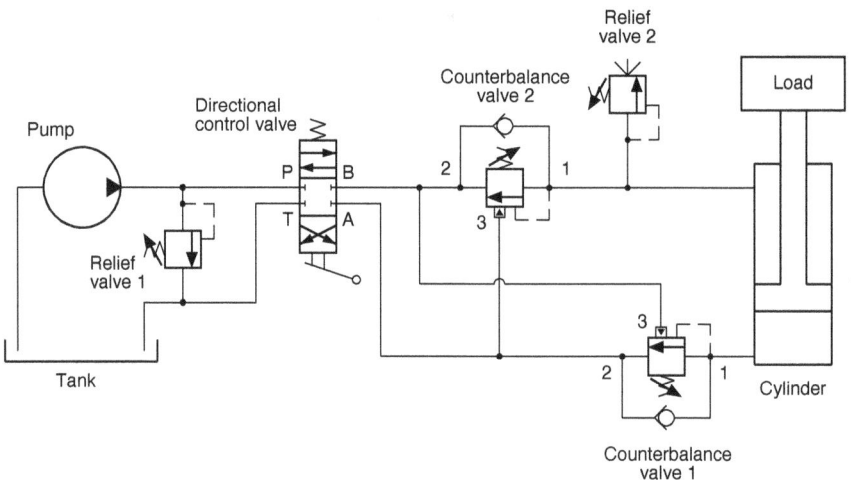

Figure 3.17 Basic Gantry Leg Hydraulic Circuit Diagram

(a)

(b)

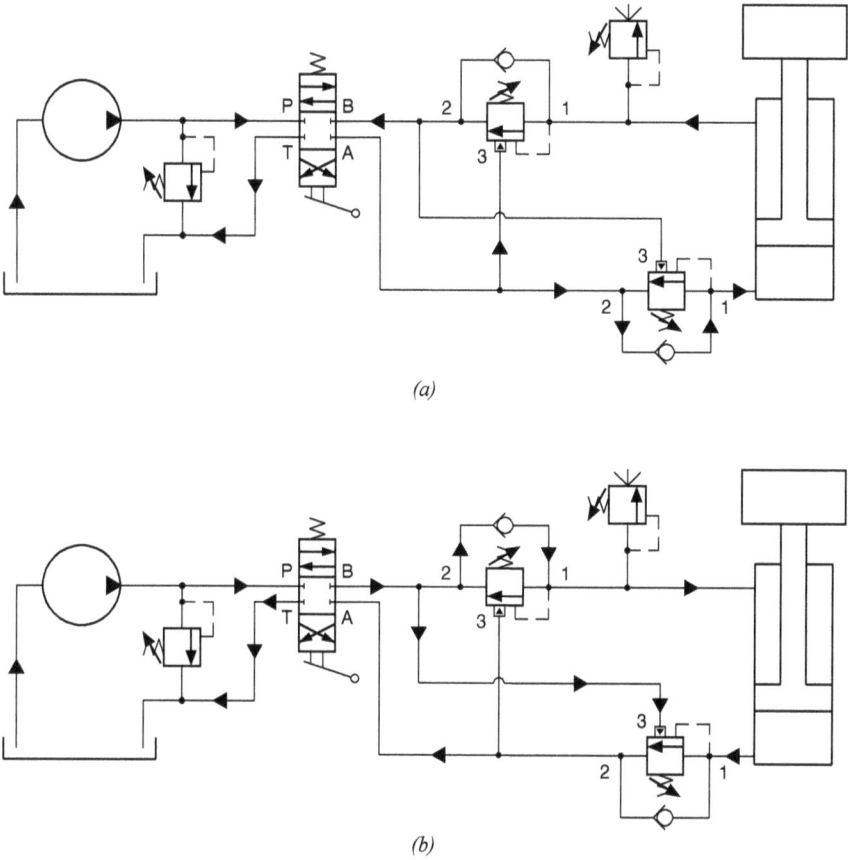

Figure 3.18 Gantry Leg Hydraulic Circuit Diagram with Oil Flow Indicated

is routed out Port T and back to the tank. Fluid flows through the check valve section of counterbalance valve 1 and into the barrel of the cylinder below the piston. As the cylinder extends, fluid is forced out of the cylinder's annulus above the piston and through counterbalance valve 2. Counterbalance valve 2 is opened by a combination of this fluid pressure at Port 1 and pressure from Port A of the directional control valve acting at the external pilot Port 3. The purpose of counterbalance valve 2 is to create a small back pressure above the piston. This back pressure assists the sleeves of a telescopic cylinder to extend, or stage, in the correct order when extending the cylinder with little or no lifted load. Last, relief valve 1 protects the system against excessive fluid pressure being delivered by the pump and relief valve 2 protects the system against a build-up of fluid pressure above the cylinder piston in the event that counterbalance valve 2 does not function properly.

To retract the cylinder, the directional control valve is shifted in the opposite direction to send fluid from the pump (Port P) to the line at Port B and to allow fluid entering the valve at Port A to exit the valve at Port T and flow back to the tank. Fluid flows through the check valve section of counterbalance valve 2 and into the cylinder above the piston. As the cylinder retracts, fluid is forced out of the cylinder barrel and through counterbalance valve 1. Counterbalance valve 1 is opened by a combination of this fluid pressure at Port 1 and pressure from Port B of the directional control valve acting at the external pilot Port 3.

Observation of the flow diagram for cylinder retraction allows us to better understand the function of counterbalance valve 1 for controlling the lowering of the lifted load. As discussed in Section 3.1.4, the cracking pressure of this counterbalance valve is set at some value greater than the pressure that will be developed at Port 1 by the maximum load that the cylinder will support. The valve will only open when a combination of increased pressure at Port 1 and a pilot pressure at Port 3 is applied. If the rate of retraction of the cylinder was to increase in an uncontrolled manner, the downward movement of the piston would reduce the fluid pressure above the piston. Since this volume is plumbed to the pilot Port 3 of counterbalance valve 1, such an uncontrolled movement would immediately allow the counterbalance valve to close, thus stopping the retraction of the cylinder. This valve function keeps the lowering of the load under the operator's control at all times.

The circuitry of a gantry leg with two or more lift cylinders will vary, depending on the intended function of the legs. The gantry system seen in Fig. 3.19 employs legs with two cylinders. However, the intended setup of these legs is such that the header beams and, therefore, the load supported by each leg will always be centered on the leg. This arrangement allows the two cylinders to share the load equally, thus permitting the two cylinders in each leg to be plumbed together with a simple tee connection. Thus, the fluid pressure in both cylinders of each pair will be equal during extension and retraction.

Figure 3.19 Gantry Legs with Two Lift Cylinders *(Lift Systems, Inc.)*

The gantry legs shown in Fig. 3.20 present a different demand on the hydraulic system. In this design, the large header plate is supported by four lift cylinders and may support two or more header beams. Since the loads on the two header beams are not necessarily equal, the loads in the four cylinders are likewise not necessarily equal.

In order to keep the header plate acceptably level, the two forward cylinders must extend and retract independently of the two rear cylinders. A level sensor on the header plate can be used to provide input to the control system to limit the out-of-level of the header plate to a pre-set maximum value. The inside and outside cylinders of each pair may be tied together, as described for the two-cylinder legs shown in Fig. 3.19. In fact, this may be desirable in order to permit the header plate to rotate slightly about its longitudinal centerline in response to header beam deflections.

3.3 HYDRAULIC SYSTEM CALCULATIONS

In this section, we will work through the calculations required to determine fluid pressures at key points in a simple gantry system that uses a single two-stage double acting cylinder on each leg. The basic hydraulic circuit to be considered is that shown in Fig. 3.17. Four cylinder positions will be investigated. These are extending the cylinder through the first stage (Fig. 3.21a), extending the cylinder through the second stage (Fig. 3.21b), retracting the cylinder through the second stage (Fig. 3.21c), and retracting the cylinder through the first stage (Fig. 3.21d). The purpose of these examples is to demonstrate the use of the equations presented in this chapter and to develop a better understanding of how the pressure readings that the gantry operator might see change as a load is lifted, held, and lowered.

Figure 3.20 Gantry Legs with Four Lift Cylinders *(Taylor Crane & Rigging, Inc.)*

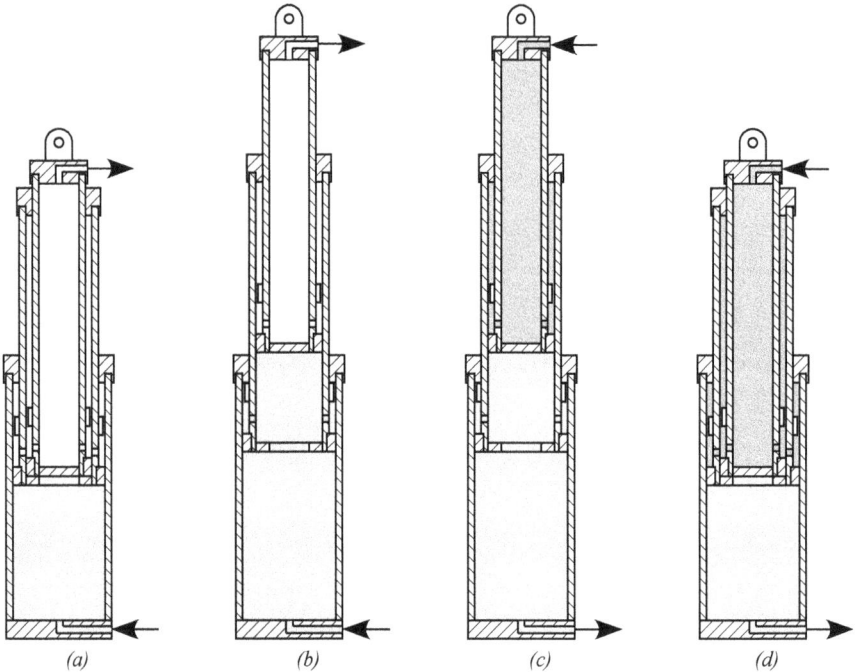

Figure 3.21 Cylinder Positions for Example Calculations

The following cylinder dimensions, load, operating pressure, and counterbalance valve settings will be used in the example calculations.

Barrel bore diameter ID_B = 12.000 inches;
Sleeve outside diameter OD_S = 11.000 inches;
Sleeve bore diameter ID_S = 10.000 inches;
Rod outside diameter OD_R = 9.000 inches;
Lifted load = 150,000 pounds;
System operating pressure = 2,600 psi;
Counterbalance valve 1 pilot ratio P_{R1} = 3.00;
Counterbalance valve 1 set (cracking) pressure S_1 = 3,250 psi (125% of the system operating pressure);
Counterbalance valve 2 pilot ratio P_{R2} = 10.00; and,
Counterbalance Valve 2 set pressure S_2 = 2,000 psi.

3.3.1   Fluid Pressure to Hold a Load

Before calculations are performed for the extension and retraction operations, it is appropriate to calculate the fluid pressures below the sleeve and rod pistons of the cylinder due to the static load. This is a very simple calculation based on the piston area and the magnitude of the supported load and will give us reference points when we examine the pressures during lifting and lowering.

EXAMPLE 3-3

Sleeve piston area $A_{B1} = \pi\ 12.000^2 / 4 = 113.10$ in.^{2}

Supported load = 150,000 pounds

First stage fluid pressure L_1 = 150,000 / 113.10 = 1,326 psi

Rod piston area $A_{B2} = \pi\ 10.000^2 / 4 = 78.54$ in.^{2}

Supported load = 150,000 pounds

Second stage fluid pressure L_2 = 150,000 / 78.54 = 1,910 psi

3.3.2   Pressures While Extending the Cylinder

When the cylinder is extended to lift the load, fluid flows unrestricted through the check valve side of counterbalance valve 1. The fluid in the annular space above the piston must pass through counterbalance valve 2 as the cylinder extends. Note that the pressure from the pump to Port 2 of counterbalance valve 1 is also the pressure applied to the pilot Port 3 of counterbalance valve 2. In the absence of live pressure at Port 1, the pilot pressure required to crack a counterbalance valve is simply the set pressure divided by the pilot ratio. Thus, we can check the state of counterbalance valve 2, as follows.

EXAMPLE 3-4

Calculate the required cracking pressure of counterbalance valve 2. Compare this pressure to the actual fluid pressure applied to the valve's pilot Port 3.

$$P = \frac{S}{P_R}$$

$$P = \frac{2,000}{10.0} = 200 \text{ psi}$$

The actual pressure at Port 3 of counterbalance valve 2 is equal to the pressure required to support the load to be lifted, which is 1,326 psi while extending the cylinder through the first stage and 1,910 psi while extending the cylinder through the second stage. This shows us that the available pilot pressure will open counterbalance valve 2, so this valve does not create a back pressure within the cylinder during extension with this load.

Example 3-4 shows us that the pilot pressure applied to counterbalance valve 2 to lift all but the lightest loads will open the valve without creating any back pressure above the pistons in the cylinder. Thus, the fluid pressure in the barrel will generally be a reasonable indicator of the magnitude of the load supported by the gantry leg.

3.3.3   Pressures While Retracting the Cylinder

Analysis of counterbalance valve 1 for retracting the cylinder is performed using Eq. 3.4. In this case, we must calculate the pressure that is required to be applied above the piston to push the cylinder down. For this calculation, we will use the following variables.

L_R = pressure applied above the piston;
L = pressure below the piston due to the supported load;
L_P = pressure below the piston due to L_R
  = L_R / C_R ; and,
L_T = total pressure applied at Port 1 of counterbalance valve 1
  = $L + L_P$.

We can also see from the hydraulic circuit diagram that the pilot pressure P to counterbalance valve 1 is equal to L_R. The use of Eq. 3.4 to calculate the required pressure above the piston to retract the cylinder and the resulting total pressure below the piston is illustrated in Example 3-5. The total pressure below the piston calculated in this example is then compared to the static pressure that was calculated in Example 3-3.

EXAMPLE 3-5

From the previous calculations, we know L = 1,910 psi in the second stage. C_R for the rod is calculated and then Eq. 3-4 is used to calculate P required for lowering through the second stage:

$$C_{R2} = \frac{\pi \, 10.000^2/4}{\pi\left(10.000^2 - 9.000^2\right)/4} = 5.26$$

$$P = \frac{3,250 - 1,910}{3.00 + 1/5.26} = 420 \text{ psi}$$

As noted above, we know that $P = L_R$, so L_R = 420 psi and

$$L_P = L_R/C_R = 420/5.26 = 80 \text{ psi}$$

Thus, the total pressure below the piston, which is also the pressure that acts at Port 1 of counterbalance valve 1, is equal to $L + L_P$ = 1,910 + 80 = 1,990 psi. This is 4.2% greater than the static pressure of 1,910 psi.

Repeating these calculations for retraction of the cylinder through the first stage yields the following results.

$$C_{R1} = \frac{\pi \, 12.000^2/4}{\pi\left(12.000^2 - 11.000^2\right)/4} = 6.26$$

$$P = \frac{3,250 - 1,326}{3.00 + 1/6.26} = 609 \text{ psi}$$

$$L_P = L_R/C_R = 609/6.26 = 97 \text{ psi}$$

Here, the total pressure below the piston is equal to $1,326 + 97 = 1,423$ psi, which is 7.3% greater than the static pressure.

For reference, we can check these results as follows. The total pressure L_T applied to Port 1 of counterbalance valve 1 is 1,990 psi in the second stage and the pilot pressure P at Port 3 is 420 psi. Thus, the effective pressure acting to crack this valve is $L_T + P_R P = 1,990 + 3.00 \times 420 = 3,250$ psi, which is the valve's set pressure S. We can perform this check for the first stage, which gives us $L_T + P_R P = 1,423 + 3.00 \times 609 = 3,250$ psi, which again is the valve's set pressure S.

The results of Example 3-5 show us that the fluid pressure measured at the cylinder barrel during the retraction of the cylinder to lower a load will be greater than the pressure due to the supported load alone. Because the percentage difference in these pressures is not a constant value for all cylinder proportions and counterbalance valve settings, determination of the supported load through the use of fluid pressure measurements is less accurate during cylinder retraction, as compared to the readings obtained during cylinder extension or the holding of the load. (From an operations point of view, once the cylinder is moving, the operator should focus on holding the pressure constant, which will show that the load on the cylinder is not changing.)

We can expand the calculations applied in Example 3-5 to show how the pilot pressure required to lower a supported load relates to the magnitude of that load. This relationship will be demonstrated for the cylinder in the second stage, but the principles apply in any stage.

We know from the example problems that the piston area in the second stage is 78.54 square inches. Thus, for a given load, the fluid pressure L to Port 1 of the counterbalance valve due to the supported load is $Load / 78.54$. The load should be specified in pounds to obtain a pressure in psi (for USCU). We have established in the input for these problems the value of the pilot ratio P_R (3.00) and we have calculated the cylinder's area ratio C_R in the second stage (5.26). We can now enter these values into Eq. 3.4 for a range of supported loads and calculate the pilot pressure P required to retract the cylinder for each load. The results of this exercise are plotted in Fig. 3.22. We see that the pilot pressure required to retract the cylinder is inversely proportional to the magnitude of the load supported by the cylinder.

This exercise and the results shown in the graph of Fig. 3.22 highlight the value of understanding how the hydraulic system functions in order to safely operate the gantry controls (for a manually operated gantry system). While the operator certainly is not expected to perform calculations like this for every lift, he must recognize that the gantry leg will not always respond in the same way to a certain

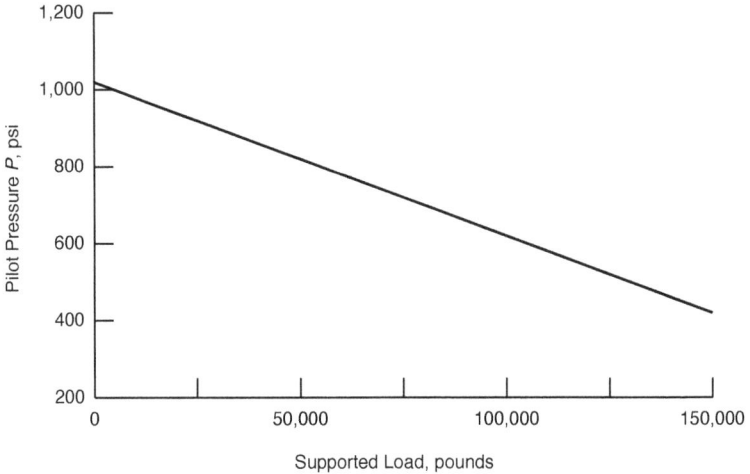

Figure 3.22 Pilot Pressure Required to Retract the Cylinder

movement of the directional control valve lever. Expanding on this, if the four legs of the typical four-leg gantry system are not loaded equally (not an uncommon occurrence), the four legs will not respond identically if all of the controls are moved identically. Rather, the more heavily loaded legs will tend to retract first since the counterbalance valves on those cylinders will open at a lower pilot pressure.

3.3.4   Extending the Unloaded Cylinder

The last function to consider is the extension of an unloaded cylinder. Of interest here is the back pressure that is developed above the piston due to the operation of counterbalance valve 2. Looking at the hydraulic circuit diagram, we see that the fluid pressure introduced at the base of the cylinder is also applied to the pilot Port 3 of counterbalance valve 2. We can modify Eq. 3.4 to provide a direct means of calculating the pressures.

The fluid pressure below the piston is equal to P and the pressure above the piston is thus equal to PC_R. This pressure above the piston is also the pressure that is applied to Port 1 of counterbalance valve 2. With these terms defined, we can now derive Eq. 3-5.

$$S = PC_R + PP_R$$

$$S = P\left(C_R + P_R\right)$$

$$P = \frac{S}{C_R + P_R} \tag{3.5}$$

Example 3-6 demonstrates the use of Eq. 3.5 to calculate the pressures acting when the unloaded cylinder is extended.

EXAMPLE 3-6

All of the values needed to solve Eq. 3.5 have been defined previously, so we can calculate the pilot pressure P directly as the cylinder extends through the first stage.

$$P = \frac{2,000}{6.26 + 10.00} = 123 \text{ psi}$$

This shows us that the external pilot pressure to counterbalance valve 2, which is also the pressure applied to the cylinder below the piston, is 123 psi. Therefore, the pressure above the piston L_R (that is, the back pressure) as the unloaded cylinder extends through the first stage is equal to $PC_R = 123 \times 6.26 = 770$ psi.

These values can be checked by noting that the pressure acting at Port 1 plus the pilot pressure multiplied by the pilot ratio must equal the valve's cracking pressure S. For this example, we find $770 + 123 \times 10 = 2,000$ psi, which is the cracking pressure of counterbalance valve 2.

Solving Eq. 3.5 for extension through the second stage gives us $P = 131$ psi and $PC_R = 131 \times 5.26 = 690$ psi.

3.3.5   Flow Rate and Volume

This last set of calculations addresses the volume of fluid that will move into and out of the cylinder as the cylinder is extended and retracted and the relationship between fluid flow rate and the speed at which the lift cylinder moves. These values are compared to the performance characteristics of the power unit.

The spaces within a hydraulic cylinder are always filled with fluid. Air and oil must never mix. Therefore, the volume of fluid that must be moved to extend or retract a cylinder is simply the swept volume through the stroke, as expressed by Eq. 3.6 for the volume below the piston and by Eq. 3.7 for the volume above the piston. In both equations, V is the volume of fluid with a subscript to define the location and L_s is the length of the stroke. The piston area A_B and annular area A_A are as previously defined by Eqs. 3.1 and 3.2.

$$V_B = A_B L_s \tag{3.6}$$

$$V_A = A_A L_s \tag{3.7}$$

Eqs. 3.6 and 3.7 are dimensionally independent. That is, the results are correct as long as all values are in the same units. Practical use of these equations calls for the cylinder dimensions (area and length) to be expressed in inches and square inches and the fluid volume to be expressed in gallons for USCU calculations or the cylinder dimensions in millimeters and square millimeters and the volume in liters for SI calculations. Eq. 3.6 is presented below with appropriate conversions for USCU calculations (Eq. 3.8a) and SI calculations (Eq. 3.8b). Eq. 3.7 can be similarly converted.

$$V_B = \frac{A_B L_s}{231} \qquad \text{(inches; gallons)} \qquad (3.8a)$$

$$V_B = \frac{A_B L_s}{1,000,000} \qquad \text{(millimeters; liters)} \qquad (3.8b)$$

We can see by comparing Eqs. 3.6 and 3.7 that the fluid flow rate through the cylinder extend port and through the cylinder retract port will not be equal, either during extension or retraction. In fact, the two flow rates will have the same relationship as the cylinder area ratio C_R. If $C_R = 4.27$, as calculated in Example 3-1, then for every gallon (or liter) of fluid that enters the cylinder below the piston during cylinder extension, $1/4.27 = 0.23$ gallon (or liter) of fluid must leave the cylinder through the retract port. The reverse is true, as well. For each gallon (or liter) that enters the cylinder through the retract port during cylinder retraction, 4.27 gallons (or liters) must leave the cylinder through the extend port. Observe how this ratio of volumes creates a hydraulic lock within the cylinder. Movement of the piston changes the total volume of fluid in the cylinder. Thus, even if fluid can leak past the piston seals, the piston cannot move unless fluid is allowed to flow into or out of the cylinder. As long as flow through both ports is blocked and the gland cap seals do not leak, the piston cannot move and the cylinder cannot extend or retract.

The relationship between the flow rate of the hydraulic fluid and the extension or retraction speed of the cylinder is expressed in a manner similar to that used to calculate the flow rates. The hydraulic pump output, which is the flow rate for the system, is identified as F_r and is most commonly expressed in gallons per minute in USCU calculations or in liters per minute in SI calculations. The cylinder extension speed E_s and the retraction speed R_s are both typically expressed in inches per minute in USCU calculations and in millimeters per minute in SI calculations. The dimensionally independent relationship is given in Eq. 3.9 for speed of extension and in Eq. 3.10 for speed of retraction (where all values must be expressed in consistent units). The units-specific relationships for speed of extension are given in Eq. 3.11a for USCU and in Eq. 3.11b for SI.

$$E_s = \frac{F_r}{A_B} \qquad (3.9)$$

$$R_s = \frac{F_r}{A_A} \qquad (3.10)$$

$$E_s = \frac{231F_r}{A_B} \qquad \text{(inches; gallons)} \qquad (3.11a)$$

$$E_s = \frac{1,000,000F_r}{A_B} \qquad \text{(millimeters; liters)} \qquad (3.11b)$$

Knowledge of the relationship between the fluid flow rate and the cylinder extension (or retraction) speed along with how the control system can function to throttle the flow rate to the cylinder can be used to evaluate the accelerations to which the lifted load will be subjected during lifting and lowering or during travel using propel cylinders. It is recognized that these flow rate calculations typically are of greatest value to a gantry system designer or manufacturer. Only in relatively rare situations will a gantry system user find it necessary to work through these numbers. Even in the absence of the need to perform these calculations, however, a sound understanding of the principles at work is still of benefit to the gantry system user and lift planner.

One example will be offered here to illustrate the utility of flow rate calculations. The gantry power unit pump is typically sized with primary consideration given to the extend and retract speeds of the lift cylinders. Example 3-7 will show the effect of unrestricted use of the available flow when applied to propel cylinders that have a much smaller bore than the lift cylinders.

EXAMPLE 3-7

Pump output rate = 40 gallons per minute (151 liters per minute)
Propel cylinder bore diameter ID_B = 3.000 inches (76.20 mm)
Number of propel cylinders being operated = 2

Maximum flow to each propel cylinder F_r = 40 / 2 = 20 gallons per minute
The cylinder piston area A_B is calculated using Eq.3.1:

$$A_B = \frac{\pi \, ID_B^2}{4} = \frac{\pi \, 3.000^2}{4} = 7.069 \text{ in.}^2$$

The cylinder extension speed E_s is then calculated using Eq. 3.11a when using USCU (as shown) or with 3.11b when using SI units:

$$E_s = \frac{231F_r}{A_B} = \frac{231(20)}{7.069} = 654 \text{ inches per minute}$$

In addition to demonstrating the use of the flow rate equations, we also see that, left unchecked, the pump output provided for acceptable lift cylinder performance can result in unacceptably fast propel or side shift cylinder movement. Rather than depend on the gantry operator to keep these movements reasonable, the gantry manufacturer can limit the extension and retraction speeds of the smaller cylinders by inserting flow restrictors in the lines to and from these components.

3.3.6   Drives for Propel and Side Shift Systems

Section 3.1 briefly introduced the use of single stage cylinders and hydraulic motors for use in gantry leg propel systems and in side shift mechanisms. It is generally not necessary for the gantry system user to perform any calculations for these items as a part of lift planning.

The gantry leg propel hardware is most commonly a part of the gantry leg as supplied by the gantry manufacturer. The performance of the propel function is generally monitored by the gantry operator in terms of the movement of the legs, not through observation of fluid pressure or flow rate. Likewise, the more complex side shifter devices are typically not designed and built by the gantry user and, again, the performance of the equipment is monitored in terms of the movement of the suspended load, not by pressure or flow readings. Thus, calculation methods for these components need not be addressed here. A brief discussion of propel system engineering is presented in Section 6.7.

3.3.7   Additional Comments

One last important point must be made: All of these calculations are wrong. Examination of the various equations presented in this section for the calculation of fluid pressures shows that all of the results are based only on the load to be moved, cylinder and valve areas, and valve pressure settings. However, hydraulic fluid does not flow through cylinders, valves, fittings, and hoses without resistance. To the contrary, every time fluid is pumped through the system, resistance is met and must be overcome.

This resistance depends on the viscosity of the fluid, the length of the hose, the radii of bends in the hose, the design of the valve or fitting, even the temperature of the fluid (which affects its viscosity). For the hydraulic system designer, there are published references, such as Crane Valves (1988), that provide extensive data on fluid flow and resistance through the components of a hydraulic system. This information can be used to refine the calculations discussed here to account for flow resistance. For the gantry user, the enhancements to the calculations required to account for this flow resistance are generally too complex to be of practical value in the realm of lift planning and execution.

The gantry lift planner and the gantry system operator must simply keep in mind that the fluid pressures calculated for the cases where the cylinder is extending or retracting are only approximate. Extending the cylinder requires overcoming friction between the cylinder components and forcing the fluid above the pistons out of the annular spaces. Resistance will be developed by this fluid and will manifest itself as an increase in the pressure below the piston needed to extend the cylinder. When the cylinder is stationary, the fluid pressures will stabilize throughout the system and will provide reasonably accurate readings. Thus, if pressure readings are used to determine the magnitude of the lifted load at each gantry leg, the most accurate measurements can be made only when the system is not moving.

This chapter provides a rudimentary discussion of hydraulic components, circuits, and calculations as applied to the hydraulic gantry system. Development of an in-depth understanding of this subject calls for further education. Two references are listed below for the reader interested in additional information on hydraulics. Daines (2009) is an introductory textbook that is comprehensive and does not presume an existing education in the field. Parker Hannifin (1993) is a more basic training manual that covers a great deal of material, but without as much technical detail as the Daines text. Many additional textbooks and handbooks are available for those in need of a higher level of understanding of this subject.

3.4 REFERENCES

Crane Valves North America (1988), *Flow of Fluids Through Valves, Fittings, and Pipe*, The Woodlands, TX.

Daines, J.R., (2009), *Fluid Power - Hydraulics and Pneumatics*, The Goodheart-Willcox Company, Inc., Tinley Park, IL.

Infrastructure Health & Safety Association (IHSA) (2007), M033 *Construction Multi-Trades Health and Safety Manual*, Mississauga, Ontario.

National Fluid Power Association, Inc. (NFPA) (2010), NFPA/T3.6.37 R1-2010 *Hydraulic Fluid Power — Cylinders — Method for Determining the Buckling Load*, Milwaukee, WI.

Parker Hannifin Corporation (1993), *Fluid Power Basics*, Cleveland, OH.

Specialized Carriers & Rigging Association (SC&RA) (1996), *Recommended Practices for Hydraulic Jacking Systems*, Centreville, VA.

4 Loads and Load Combinations

The design of a telescopic hydraulic gantry system and the individual components that make up the system, as well as the engineering of a lift using a gantry system, must be based on a sound understanding of the functions of the equipment and the loads that are imposed upon the equipment as a result of the performance of those functions. When attempting to quantify those loads, it is tempting to simply adopt provisions from existing design standards that apply to other types of lifting equipment. Although a gantry system may perform the same basic work as a mobile or overhead crane, its operation is different enough from that of a crane that crane design and setup practices do not necessarily apply to gantries. However, as with a mobile or overhead crane, the loads imposed upon the gantry system during a lift can cause a structural or mechanical failure or a loss of stability if not adequately accounted for in the engineering and operation of the gantry. Both designers of gantry products and the users of the systems must understand the nature of these loads and how they affect the performance of the gantry system.

This chapter examines the various types of loads that act on a gantry system during a lifting operation. Methods by which these loads can be calculated are presented, augmented by practical guidelines for design applications. Assumptions are discussed where appropriate.

4.1 THE LIFTED LOAD AND OPERATIONAL LOAD EFFECTS

The structural and mechanical components of a hydraulic gantry system must be designed for loads that may be divided into six categories. These are:

- Gravity loads due to the weights of the object being lifted and any associated hardware (slings, shackles, etc.).
- The load effect due to unequal extension or retraction of the gantry legs, called cross cornering.
- Forces created by misalignments of the components of the lifting system, such as picking the load off-center or drifting the load when setting it.
- Dynamic loads, acting either vertically or horizontally, that are created by the motion of the lifted load and/or the equipment.
- Forces that are developed when the gantry system legs on one side become misaligned with the legs on the other side during travel, called racking.

- Environmental loads, which for mobile lifting equipment usually means wind loading. In some special circumstances, however, the consideration of seismic loads may be necessary.

In addition to determining the magnitudes of these various loads and forces, the engineer must also understand how these loads and forces act together. It is intuitively obvious that all of the loads and forces do not act simultaneously at their maximum values. Thus, developing rational load combinations is an important facet of the design process.

4.1.1 Gravity Load

The gravity load acting on a hydraulic gantry system is often very well defined. For the designer of the gantry system, the gravity load is the specified rated load of the gantry leg or system component. The manufacturer's product design engineer may reasonably use this rated load as the design basis, thus creating a responsibility on the part of the gantry system user not to exceed the rated load. This approach is, of course, consistent with the design of and use responsibilities for all lifting equipment.

For the gantry system user, the gravity load is the sum of the weights of the items being lifted. The weight of the payload being handled can generally be determined with reasonable accuracy. The weights of all of the rigging components, from the shackles that attach to the payload's lifting lugs to the header beams that span between the gantry legs, are usually well known. All of these weights are resolved into reactions from the header beams to the gantry legs as a part of the header beam design (header beam design is addressed in Chapter 6). Finally, the computed gravity load applied to each leg can be verified, at least approximately, as the lift begins by reading the hydraulic pressures in the lift cylinders as the load is raised.

4.1.2 Cross Cornering

Many of the items commonly lifted with hydraulic gantries, such as presses and power generation components, are rigged to four lift points. The portion of the total lifted weight carried at each lift point is typically calculated by considering only the geometry of the lift point arrangement and the location of the center of gravity using Eq. 4.1.

$$R_n = \frac{P}{N} \pm \frac{M_x y}{I_x} \pm \frac{M_y x}{I_y} \tag{4.1}$$

where
R_n = reaction at lift point n;
P = total weight of lifted load;

N	=	number of lift points;
M_x	=	moment about the transverse axis due to the longitudinal offset of the center of gravity;
I_x	=	moment of inertia of the lift points about the transverse axis;
y	=	longitudinal distance from the transverse axis to lift point n;
M_y	=	moment about the longitudinal axis due to the transverse offset of the center of gravity;
I_y	=	moment of inertia of the lift points about the longitudinal axis; and,
x	=	transverse distance from the longitudinal axis to lift point n.

Eq. 4.1 can be simplified to Eq. 4.2 for a lifted load with four lift points arranged on the corners of a rectangle.

$$R_n = \frac{P}{4} \pm \frac{M_x}{S_x} \pm \frac{M_y}{S_y} \qquad (4.2)$$

where

S_x	=	section modulus of the lift points about the transverse axis;
	=	$4 (L/2)^2/(L/2) = 2L$ for a rectangular arrangement;
S_y	=	section modulus of the lift points about the longitudinal axis;
	=	$4 (B/2)^2/(B/2) = 2B$ for a rectangular arrangement; and,
		all other terms are as previously defined.

EXAMPLE **4-1**

Consider the item shown graphically in Fig. 4.1. The item is lifted from four lift points and is of substantial torsional stiffness. Determine the four corner reactions of the lifted load shown in using Eq. 4.2.

Eq. 4.2 can be written as follows with the appropriate section modulus values and the moments expressed as functions of P.

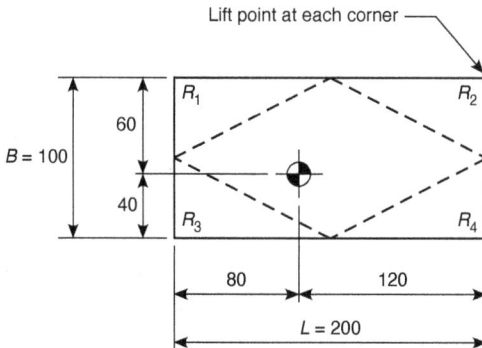

Figure 4.1 Plan View of a Load Lifted from Four Points

$$R_n = \frac{P}{4} \pm \frac{P(200/2-80)}{2(200)} \pm \frac{P(100/2-40)}{2(100)}$$

Solving Eq. 4.2 for $P = 100$ gives us $R_1 = 25$, $R_2 = 15$, $R_3 = 35$, and $R_4 = 25$.

In reality, this is a statically indeterminate problem. Thus, the true load at each of the four lift points will be affected by the torsional stiffness of the item and the axial stiffness of the rigging at each support. This load distribution may actually occur during the lift, but only if the rigging lengths, equipment setup, and gantry system operation are all nearly ideal. This usually isn't the case.

Note that Eqs. 4.1 and 4.2 will only produce valid results for four lift point reactions when the center of gravity is located within the area defined by the dashed lines in Fig. 4.1. If the center of gravity is outside of this area, the reaction at the corner diagonally opposite the quadrant in which the center of gravity falls will become negative. If the load is suspended from slings or other types of rigging that can only carry a tension load, a negative support reaction is not physically possible. Rather, that sling will go slack and the weight will be carried entirely by the remaining three slings. Calculation of the lift point reactions is now a statically determinate problem for which the results can be reliably obtained using basic statics.

It is probable that the actual load distribution among the four lift points will vary from that predicted by Eqs. 4.1 and 4.2, primarily due to the operation of the gantry system. When lifting a very rigid load from four lift points, the sharing of the load among the four points may be altered from the theoretical distribution by uneven extension of the gantry legs. This effect is called cross cornering.

The load distribution among the four lift points is affected by the magnitude of the difference of leg extensions, the bending stiffness of the header beams, the axial stiffness of the rigging, the spread of the lift points as compared to the span of the header beams, and the offset of the lift points with respect to the center of the header beam span. We can best examine this behavior by means of a few examples.

EXAMPLE 4-2

Consider the lift illustrated in Fig. 4.2. The load weighs 400,000 pounds (181,437 kg) and its four lifting lugs are symmetrically located about the center of gravity. The header beams are W14x311 wide flange members that span 20 feet (6.10 meters), the load is centered among the gantry legs in both horizontal directions, and the slings are wire rope with a diameter of 2 1/2" (65 mm) and a length of 24 feet (7.32 meters), each rigged in a basket hitch. The lift links are 10'-0" (3.05 meters) center to center and the header beams are 12'-0" (3.66 meters) center to center. (Note that Fig. 4.2 is not drawn to scale.)

If the four gantry legs extend in perfect unison, the load will be carried equally among the four slings and the four legs, as expected. If one leg should extend a little faster or a little slower than the others, this balance will be upset. Assume that Gantry Legs 1, 2, and 3 extend in unison but that Gantry Leg 4 lags the other three legs by about 1 1/2" (38 mm). Instead of

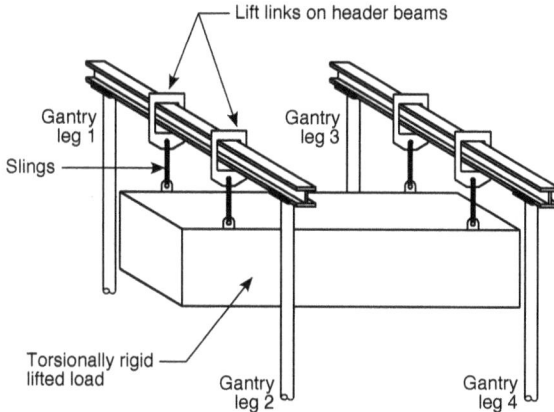

Figure 4.2 Lifting a Torsionally Rigid Load

each sling carrying 100,000 pounds (45,359 kg), the slings nearest to Gantry Legs 2 and 3 will each carry 200,000 pounds (90,718 kg) and the slings nearest to Gantry Legs 1 and 4 will go slack. As a result of the redistribution of sling tensions, the bending moment in the header beams due to the vertical load will increase by 50%. The loads in Gantry Legs 1 and 4 due to the weight of the lifted item will be about 50,000 pounds (22,680 kg) each and the loads in Gantry Legs 2 and 3 will be about 150,000 pounds (68,039 kg) each.

EXAMPLE 4-3

We can repeat the analysis of Example 4-2, but with the center of gravity of the lifted item offset 30 inches (762 mm) in both horizontal directions toward Gantry Leg 4. All other characteristics of the problem remain as described in Example 4-2. In this case, a lag at Gantry Leg 4 of only 1/8 inch (3 mm) will result in the sling nearest to Gantry Leg 1 going slack and the load then being carried by the remaining three slings. (Since the three-point load distribution is a statically determinate problem, further loss of synchronicity in leg extension will not increase the load imbalance.) The maximum gantry leg load calculated using the lift point loads determined using Eq. 4.1 and standard header beam design methods gives us a maximum gantry leg load of 166,667 pounds (75,599 kg) at Gantry Leg 4. The load at Gantry Leg 4 is only 162,500 pounds (73,709 kg) at the point where the sling at Gantry Leg 1 goes slack. The load in Gantry Leg 1 drops slightly and the loads in Legs 2 and 3 increase, but the load at Gantry Leg 4 remains the maximum. Because of this change in gantry legs loads, the maximum bending moment in the more heavily loaded header beam drops by about 2 1/2%.

EXAMPLE 4-4

For this third example, we will take the asymmetric load of Example 4-2 and offset it on the header beams by 30 inches (762 mm) toward Gantry Legs 1 and 3. As with Example 4-2, a difference at Gantry Leg 4 of about 1/8 inch (3 mm) results in the sling nearest to Gantry Leg 1 going slack, again making the remaining three sling tensions statically determinate. The values of the sling tensions and gantry leg reactions are somewhat different due to the different header beam deflections at the lift links, but the overall behavior is the same.

The magnitude of the load imbalance due to cross cornering and the gantry leg extension difference at which the maximum effect is observed are functions of the relative stiffnesses of the components involved (the lifted item, the rigging, and the header beams) and the geometry of the load and the gantry system. The examples shown assume that the lifted item is very rigid torsionally. The cross cornering effect will be reduced if the lifted item can flex in response to changing rigging tensions. The cross cornering effect can also be reduced by using slings of lesser stiffness. Consider the two lifts shown in Fig. 4.3. Both loads, a press crown in Fig. 4.3*a* and a transformer in Fig. 4.3*b*, are very stiff torsionally. The press crown is rigged with four short wire rope slings in basket hitches. These slings are very stiff axially. The transformer is rigged with four synthetic slings which are much more elastic than wire rope slings of similar length and rated load. Thus, if all other aspects of these two lifts were identical, the press crown lift with the wire rope slings would be much more susceptible to loading imbalance due to cross cornering than would be the transformer lift.

With respect to load geometry, the worst case occurs when the center of gravity of the lifted item is centered among the four lift points, the lift links are widely spaced on the header beams, and the lift links are centered on the beams. This arrangement offers optimum load distribution if all aspects of the lift are ideal but can result in a 100% / 0% split of the load between pairs of diagonally opposite corners. The wider the spread of the lift links on the header beam, the greater will be the change in the reactions to the more heavily loaded gantry legs. This, in turn, produces greater changes in the major axis bending moment and shear in the header beams.

The upper bound increase in the lift link load for a rectangular item lifted at four points is 100%, which occurs when a symmetrical item is carried by only two diagonally opposite lift links. The upper bound increase in the header beam reaction to a gantry leg at full cross cornering at the lift links is on the order of 90%

Figure 4.3 *(a)* Lifting with Wire Rope Slings. *(b)* Lifting with Synthetic Slings.
(Burkhalter Rigging, Inc.)

and occurs with a symmetrical load, symmetrically placed lift links, and a wide lift link spread. The results of a series of analyses of fully symmetrical lifts are plotted in Fig. 4.4. We see here that the upper bound percentage increase in the gantry leg load is equal to the ratio of the lift link spread divided by the header beam span, expressed as a percentage.

Note that the gantry leg extension difference at which the upper bound loading occurs is inversely proportional to the ratio of the load point spread to the span. That is, the wider the lift link spread, the more sensitive the system is to cross cornering, and the less extension difference is needed to reach full cross cornering. This effect is illustrated in Fig. 4.5 for three different ratios of lift link spread to header beam span. The increase in the split of the load between diagonally opposite pairs of gantry legs increases linearly until full cross cornering of the load has been reached. At this point, the load is carried entirely by two diagonally opposite slings and further deviation of the gantry leg extensions will have not further effect on load distribution. These points are indicated in Fig. 4.5 by the change in the plotted lines from sloped to horizontal.

The calculations that underlie the curves in the graph of Fig. 4.5 are based on the load, rigging, header beam properties, and gantry leg layout defined in Example 4-2. The lift link locations on the header beams remain symmetrical, with only the spread differing, as indicated in the figure. Calculations performed using different input values will yield different numerical results. However, the general trends seen in Fig. 4.5 will hold.

Clearly, cross cornering can result in load distributions that are significantly different than those based on the assumption of balanced loading. This behavior affects the rigging, the header beams, the cross beams, if applicable, and the gantry legs. Also clearly, however, cross cornering can be minimized by careful monitoring and operation of the system. The examples above show that the gantry

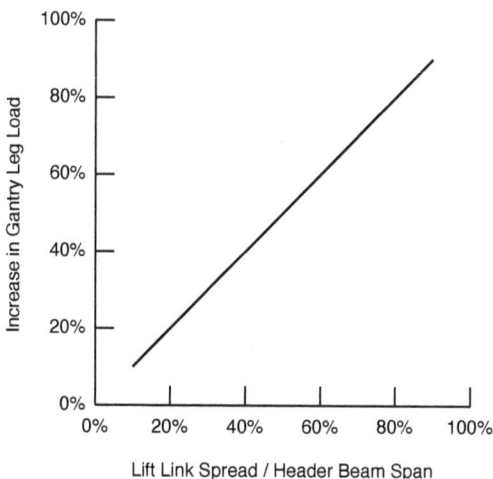

Figure 4.4 Effect of Load Point Spread on Increase in Gantry Leg Load

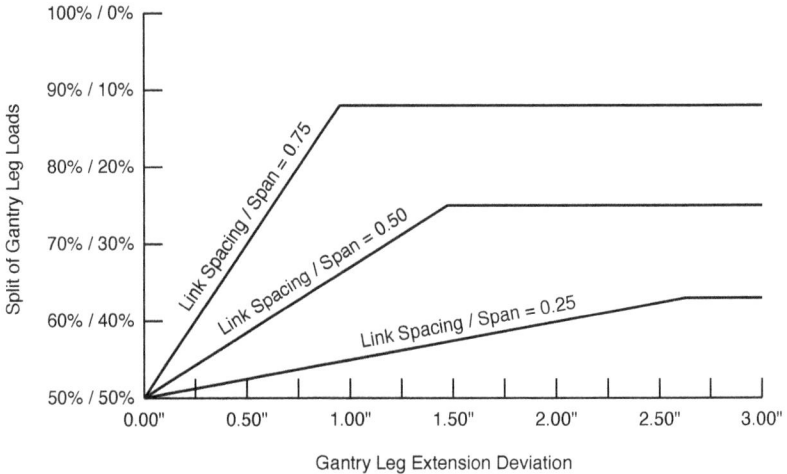

Figure 4.5 Effect of Lift Link Spacing on Cross Cornering

operator can minimize cross cornering in any of three ways. First, watching the hydraulic pressure gauges or other load indicating devices on the control unit will allow the operator to maintain consistent loading in each of the legs. Second, the use of leg extension measuring devices (which may be as simple as steel tape measures fixed to each leg) can be used to assure uniform leg extension and retraction as the lift proceeds. Third, the use of an automated control system that continually monitors load and extension at each leg and adjusts the leg movement automatically when preset tolerances are exceeded will minimize cross cornering.

One aspect of the three examples detailed above is that the four slings are exactly the same length. This is the ideal, but not always realized. If the two slings rigged to one header beam are slightly different in length [say a difference in length of no more than an inch (25 mm) or so], the operator may be able to compensate by extending one gantry leg a little more than the other to snug up both slings before beginning the lift. This may not always be a practical solution, however, since it is generally desirable to keep the header beams level throughout a lift. To continue with our example, if the lift links are 10'-0" (3.05 meters) apart and the sling lengths vary by 1 inch (25 mm), tightening the slings by allowing the header beam to be out of level will require a header beam slope of 0.83% (0.48°). This is not particularly severe and may be acceptable in some cases.

4.1.3   Misaligned Rigging and Out-of-Level Setup

The primary non-inertia-related horizontal load effect that must be consider is the horizontal component of the tension in the rigging that occurs when the rigging is not precisely vertical. When initially lifting a load, such out-of-plumb rigging is most likely an unplanned condition (that is, picking a load with the lift links purposely misaligned is generally not considered an acceptable practice). When upending or laying down a load, while out-of-plumb rigging is still not desirable, the condition is created by the operation and must realistically be considered unavoidable. Last, it is sometimes necessary to drift the suspended load horizontally to align base plate holes with anchor bolts. This need occurs when the gantry system is slightly out of alignment with the foundation of the item being lifted.

This effect is illustrated in the lateral direction in Fig. 4.6. The lift points on the load to be lifted are offset with respect to the lift links. This results in the slings being out of plumb as the load is initially lifted. The angle of the rigging with respect to vertical creates a horizontal force at each lift link, as illustrated in the figure. This horizontal force is equal to $T \tan \theta$, where T is the design rigging tension based on plumb rigging and θ is the angle of the rigging from vertical, or $T s$, where s is the slope of the rigging from vertical, in percent.

In addition to the side pull as a misaligned load is first lifted, once the load is free, it will sway (Fig. 4.7). Although the effect acting on the gantry system is now a dynamic effect, we can see that the inertial force from the swinging load will not be any greater than the horizontal component of the rigging tension due to the initial misalignment. Therefore, this load effect is considered to be due to the misaligned rigging and not the dynamic effect.

A rigging arrangement is generally not surveyed with instruments. Rather, the condition of plumb, or the lack thereof, is estimated by eye. Based on common lengths of slings, shackles, links, and other rigging components typically used when lifting with a gantry system, the length of rigging between the header beam and the lifted load will not normally be less than about 3 feet (1 meter). A slope of 1.5% relative to vertical equates to an out-of-plumb of a little over 1/2 inch

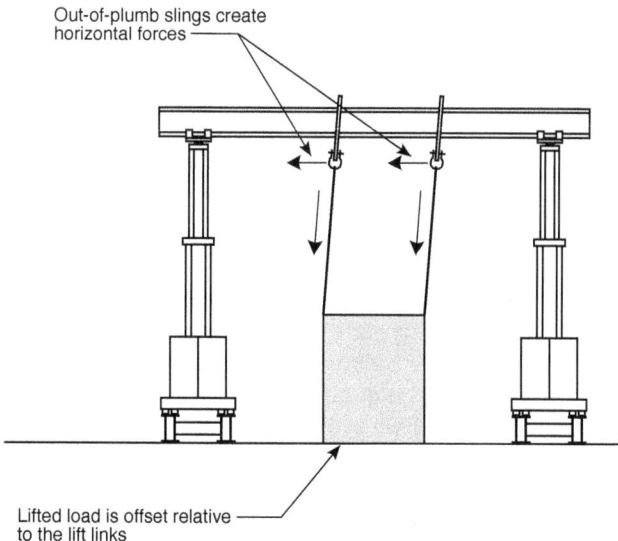

Out-of-plumb slings create horizontal forces

Lifted load is offset relative to the lift links

Figure 4.6 Load Misaligned Relative to the Lift Links

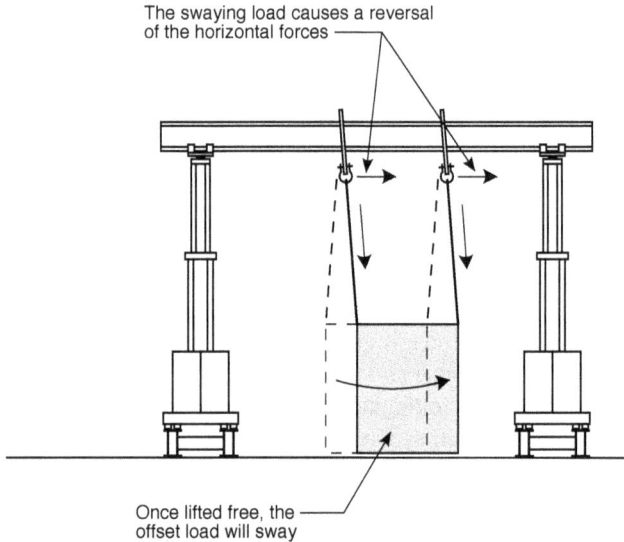

The swaying load causes a reversal of the horizontal forces ⎯⎯

Once lifted free, the ⎯⎯ offset load will sway

Figure 4.7 Motion Resulting from Lifting a Misaligned Load

over 3 feet (15 mm in 1 meter). It is reasonable to expect a worker to be able to view such a dimension by sighting along a building column, the gantry leg, itself, or another object that is known to be plumb. This indicates that we may assume that a reasonable maximum out-of-plumb of the rigging is about 1.5% and, correspondingly, the maximum horizontal force that will act on the gantry system due to this out-of-plumb rigging is similarly equal to 1.5% of the weight of the suspended load. Last, although this condition is illustrated in the lateral direction, the same effect can occur in the longitudinal direction.

The drifting of a suspended load for anchor bolt alignment should be minimal [no more than an inch (25 mm) or so]. The need for drifting arises from a misalignment of the gantry system when set up and can be minimized through care when laying out and installing the tracks.

We can see a relationship between the length of the rigging and the out-of-plumb effect. The longer the rigging, the smaller will be its angle off vertical for a given misalignment. Whether considering the detection of an initial misalignment of the lift links relative to the lift points or the intentional misalignment that occurs when drifting a suspended load, longer rigging reduces the resulting horizontal loading. However, the use of longer rigging will require that the gantry legs be extended to a greater height for a given lift. This has a detrimental effect on the stability of the gantry system, as is discussed in Chapter 5. These two effects must be balanced against one another when planning a lift.

It is common practice to allow one pair of gantry legs (in a four-leg arrangement) to freewheel when initially lifting a load. This allows the system to move longitudinally in response to the horizontal pull of out-of-plumb rigging to center itself around the load. This tells us that the maximum horizontal load that can act on the gantry system in the longitudinal direction due to out-of-plumb rigging is that force required to overcome the rolling resistance of the gantry leg wheels on the track.

We need two values to assess the longitudinal horizontal load as limited by the rolling resistance of a gantry system. The first value is, of course, the rolling resistance multiplier typical to a gantry leg. The second value is the ratio of gantry leg self weight to rated load, since it is most convenient to express the horizontal load as a percentage of the vertical load supported by the leg.

One of the gantry system manufacturers has performed rolling resistance tests on a gantry system with an external propel device and measured the resistance to be approximately 1.3% of the total vertical load. This value is reasonably consistent with the value of 0.97% measured in tests of industrial rollers conducted by Lehigh University in 1991. (Descriptions and results of these two test programs have not been published formally, but have been widely circulated among lift engineering specialists.)

The self weight of a gantry leg expressed as a percentage of the rated load at maximum extension varies significantly. The range is about 4% to 11% for bare cylinder models and 7% to 10% for telescopic boom models, based on current production models. For our purposes here, we can apply a value of 10% for all models to get a reasonable upper bound load multiplier for rolling resistance.

If the rolling resistance of a leg is 1.3% of the total vertical load and the self weight W_d is equal to 10% of the rated load RL, then we can express the rolling resistance as 1.3% x $(RL + W_d)$ = 1.3% $(RL + 0.1\ RL)$ = 1.3% x 1.10 RL = 1.43% of the rated load. We can round this result up to 1.5% for practical applications of equipment design and lift planning.

Out-of-plumb rigging can occur in the longitudinal direction when using a gantry system to upend or lay down a load (Fig. 4.8). Here, too, it is accepted practice to allow one pair of gantry legs to freewheel when upending or laying over

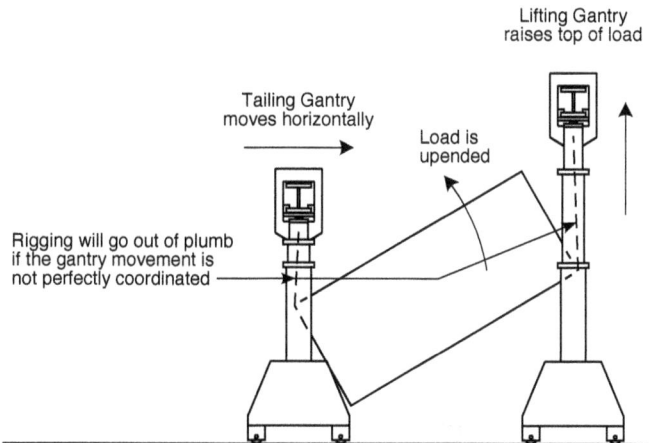

Figure 4.8 Out-of-Plumb Rigging due to Upending a Load

a load. Thus, the same upper bound longitudinal horizontal force discussed above (1.5% of the vertical load) can be applied to this condition.

The rigging can also be forced out of plumb in the longitudinal direction by uncoordinated travel of the legs. Consider a four-leg gantry system as shown in Fig. 4.9. If the rigging is plumb and all four legs travel uniformly, the rigging will remain plumb. However, if one set of two legs moves ahead of or behind the other set, the slings will be forced out of plumb. This effect can be controlled either by monitoring the spacing of the gantry legs as the system travels or by observing the rigging relative to an object of known plumb, such as the gantry legs themselves. The upper bound of this load effect can be assumed to be the same as that for initial misalignment: 1.5%, as limited by observational precision.

The loads acting on the gantry leg are generally considered to be acting coaxial and perpendicular to the centerline of the leg. This, of course, is also the most convenient means of applying the loads to the leg for the analysis of the leg. This geometry is upset when the leg is not plumb due to the supporting surface being out of level. The effect of an out-of-level base can most easily be handled by resolving the applied vertical and horizontal forces into the coordinate system in which the gantry leg is vertical.

Consider the case shown in Fig. 4.10. The gantry leg is out of plumb as a result of an out-of-level base. The applied vertical and horizontal loads are P and F, respectively, and the angle by which the base, and therefore the leg, is out of level is ϕ. The figure illustrates the conversion of the actual loads into a pair of loads that are axial with and transverse to the gantry leg. The resolved axial and transverse loads P_r and F_r are computed using Eqs. 4.3 and 4.4.

$$P_r = P\cos\phi - F\sin\phi \qquad (4.3)$$

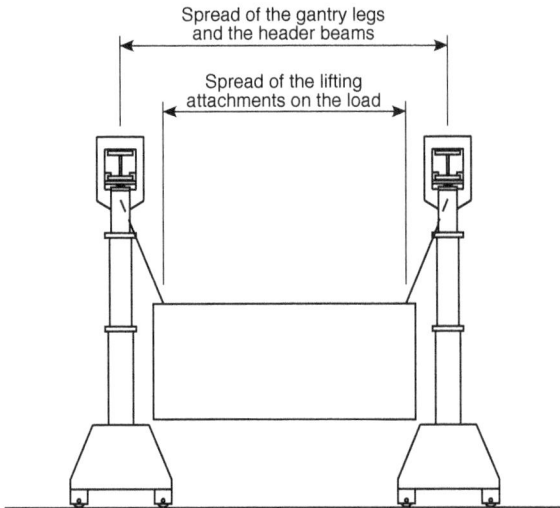

Figure 4.9 Out-of-Plumb Rigging Caused by Uncoordinated Travel

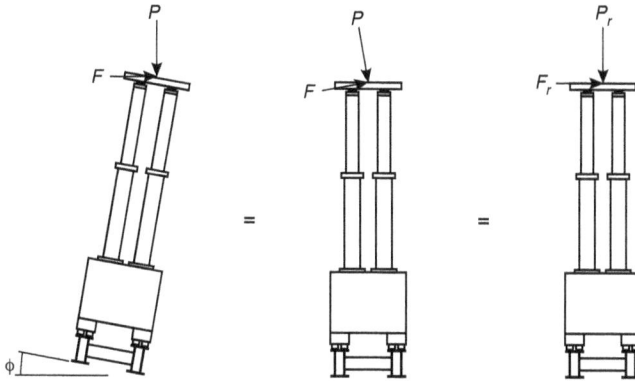

Figure 4.10 Resolution of Loads with an Out-of-Level Base

$$F_r = P \sin\phi + F \cos\phi \qquad (4.4)$$

Note that the true coordinate system of the gantry legs must be used in stability calculations (refer to Chapter 5 for details of how an out-of-level base affects stability).

4.1.4 Dynamic (Inertial) Loads

The fourth group, motion-related loads, bears careful investigation. This group of loads includes dynamic (inertial) loads and load effects acting on the gantry system due to movement of the payload during the lifting operation and due to movement of the overall gantry system.

Dynamic forces are created by the inertia of a mass as it accelerates from rest or decelerates from motion. For our purposes here, we must consider dynamic forces that act vertically and those that act horizontally. Vertical dynamic loading may be caused the acceleration and deceleration that occurs when starting and stopping a lift, by vertical oscillation (bounce) of the lifted load due to the elasticity of the header beams, slings and other rigging during any movement of the load, or by any sudden shifting of the load during the lift, as may occur during the laying over of a load.

Horizontal dynamic loads are created by similar events, but are particularly pronounced in operations that involve rolling the gantry system along its track with the load suspended or when upending or laying over a load. Rolling the gantry system with the load suspended will create a condition in which the load will sway slightly, thus inducing a horizontal load in the longitudinal direction. Laterally moving the suspended load along the header beams (side shifting) can similarly result in a swaying of the load with the corresponding creation of a horizontal load in the lateral direction. Due to the most common proportions of gantry legs, the system is typically most sensitive to lateral loads.

Vertical Inertial Loads.   The vertical movement of the load when lifting and when lowering is normally very slow, smooth, and well controlled. The hydraulic controls of a gantry system provide the operator with variable control of fluid flow (unlike, for example, some electric overhead cranes that have simple on-off switches to control movement of the load). Further, the automated control systems on many newer gantry models limit the acceleration and deceleration rates to provide slow and predictable motions. Thus, there is an absence of abrupt starting and stopping of the extend and retract motions of the gantry legs. We may consider inertial loading due to extension and retraction of the gantry legs to be nominally zero for normal gantry operation.

Horizontal movement of the suspended load along the header beams and travel of the gantry system along its track will develop an increase in the vertical load acting on the header beams or track beams, respectively. Consider a simplified version of this problem. When a weight W moves across a simply supported beam where the weight of the beam is very small relative to weight W, dynamic loading can be determined as a function of the maximum ratio of the dynamic deflection Δ_d of the beam to its static deflection Δ_s. This ratio is calculated using Eq. 4.5 (Young, et al 2012).

$$\frac{\Delta_d}{\Delta_s} = 1 + \frac{v^2}{g}\frac{WL}{3EI} \qquad (4.5)$$

where

v	=	speed at which weight W moves;
g	=	acceleration due to gravity;
L	=	beam span;
E	=	modulus of elasticity of the beam material; and,
I	=	moment of inertia of the beam.

Eq. 4.5 is dimensionally independent, so values in any consistent set of units can be used for solution of a particular problem.

We can illustrate with a simple example the influence of this behavior using values that are appropriate (or typical) with respect to gantry system use and performance.

EXAMPLE 4-5

Consider a gantry system header beam as described below supporting the indicated load that moves along the beam at the stated speed:

W = 100,000 pounds

L = 12 feet = 144 inches

v = 7.5 feet per minute = 1.5 inches per second

g = 386.1 inches / sec^2

E = 29,000,000 psi

I = 1,530 in.^{4} (a W14x132 rolled shape)

Entering these values into Eq. 4.5, we compute the ratio of the dynamic to static deflection of the header beam.

$$\frac{\Delta_d}{\Delta_s} = 1 + \frac{1.5^2}{386.13} \frac{100,000(144)}{(29,000,000)1,530} = 1.000001$$

The result shown in this example demonstrates that the dynamic amplification due to horizontal movement of either the load on the header or cross beams or the movement of the gantry system along track beams is negligible. The selected weight, beam span, and beam size are all such that the beam is acceptable for this use, based on common gantry lift planning practices. We can see that changes to the problem input, such as increasing the travel speed v or decreasing the beam stiffness (by changing either E or I) within practical ranges will not be significant enough to alter the conclusion of the example.

A true analysis of the dynamic amplification in a gantry system must consider the elasticity of the rigging between the load and the header beams, the header beams, the gantry legs, and the supporting surface. These elasticity values coupled together will yield a total stiffness of the support of the moving load that is less than that calculated using Eq. 4.5 and simply the beam stiffness. Regardless of the final stiffness value, we can see that the side shift and propel speeds common to gantry systems are low enough that a change in the support stiffness of even two or three orders of magnitude will still lead to the conclusion that the vertical dynamic loading due to the horizontal motion of or within a gantry system is trivial.

One potentially significant source of vertical impact is the jolt that can occur when using gantries to lay down a load (the reverse of the operation illustrated in Fig. 4.8). When planned correctly, the load distribution to the gantry system will be known throughout the movement and the motion will be smooth and controlled. However, an error in the determination of the center of gravity of the load can inadvertently result in unexpected movement of the load or a distribution of weight between gantries that varies from the computed distribution. Such an operation should absolutely never be planned with the intention of letting the load roll over center and then "catching" the load with the gantries. Because the dynamic loading from this type of operation is entirely indeterminate, one cannot realistically compute the magnitude of the load. The assumption of a suitable value for lift planning must be made on the basis of engineering judgment on a case-by-case basis.

Chapter 2 discusses of the use of gantry systems with other types of lifting equipment, particularly an underhung trolley on the header beam (Fig. 4.11) and strand jacks (Fig. 4.12). In these arrangements, the lifting characteristics of the integrated equipment can develop dynamic loading unlike that typical to gantry operation.

Let us consider first the underhung trolley. The conventional design and installation of an underhung trolley and hoist is addressed by CMAA (2010). This specification defines dynamic load factors for use in the design of this equipment.

Figure 4.11 Lifting Using an Underhung Trolley and Hoist *(J&R Engineering Co., Inc.)*

Two vertical load factors are defined, one that is applied to the dead load of the trolley and hoist and one that is applied to the lifted load. The dead load factor *DLF* is given in Eqs. 4.6*a* and 4.6*b*, where v_t is the trolley travel speed in feet per minute (Eq. 4.6*a*) or in meters per minute (Eq. 4.6*b*). The hoist load factor *HLF* is given in Eqs. 4.7*a* and 4.7*b*, where v_h is the hoist speed, again in feet per minute (Eq. 4.7*a*) or in meters per minute (Eq. 4.7*b*). (The reader should note that these two equations are simplified expressions and do not rigorously compute the true dynamic loads that will be imposed upon the system during a lift.)

$$DLF = 1.1 \leq 1.05 + \frac{v_t}{2000} \leq 1.2 \qquad (4.6a)$$

$$DLF = 1.1 \leq 1.05 + \frac{v_t}{610} \leq 1.2 \qquad (4.6b)$$

Figure 4.12 Strand Jacks Mounted on a Gantry System *(Burkhalter Rigging, Inc.)*

$$HLF = 1.15 \leq 1 + 0.005 \ v_h \leq 1.5 \qquad (4.7a)$$

$$HLF = 1.15 \leq 1 + 0.0165 \ v_h \leq 1.5 \qquad (4.7b)$$

The load factors defined by Eqs. 4.6a and 4.7a are graphed in Fig. 4.13. We can see in the figure that the calculated values of both of the load factors, in the absence of the specification-imposed minimum values, would likely be trivial for trolley travel speeds that are practical for gantry applications. Thus, the values given by Eqs. 4.6 and 4.7 should be regarded as required for compliance with a specification and not representative of true dynamic loading that will be imposed on the gantry system. This conclusion is consistent with the finding of Example 4-5.

The applicability of the Hoist Load Factor depends on how the overall system is operated to make the lift. Consider the gantry system illustrated in Fig. 4.11. If the Liberty Bell is lifted and lowered by means of the chain hoist, then the use of the CMAA (2010) Hoist Load Factor may be appropriate to demonstrate compliance with an accepted industry standard. If the Bell is lifted and lowered by extending and retracting the gantry legs, then the chain hoist is functioning as simple rigging and the previous discussion about vertical inertial loads from the gantry leg extension and retraction motion applies.

The use of strand jacks is not as well defined. There are presently no industry standards or specifications that define engineering requirements for lifting with strand jacks. Therefore, it remains the sole responsibility of the lift planner/engineer to determine suitable dynamic load factors to apply in the design of such a setup. It is noted, however, that the vertical movement of strand jacks, like that of hydraulic gantries, is exceptionally slow and that the true dynamic loading due to lifting with strand jacks is also near zero.

As of this writing, a new chapter is under development for ASME B30.1 (ASME 2009) in which lifting with strand jacks will be addressed. It is uncertain at this time what dynamic load factors, if any, will be recommended or required by

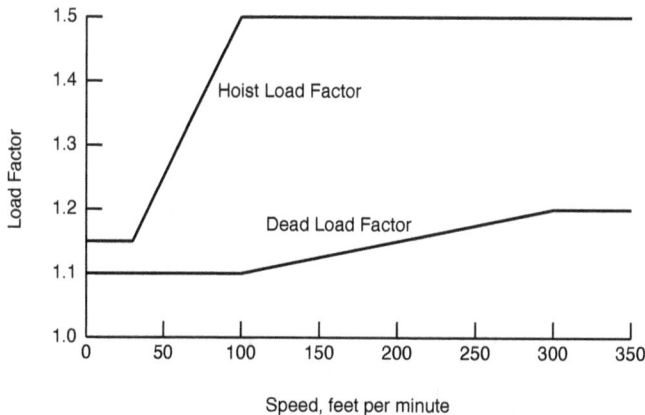

Figure 4.13 CMAA (2010) Underhung Crane Load Factors

the final version of this volume. It is expected that the next edition of B30.1 will be published in 2014.

Last, it is common practice in rigging planning and engineering to increase the gravity load by some multiplier to account for possible inaccuracies in the stated weight of the item being lifted. These multipliers are often called impact factors. In reality, these multipliers are being used to address uncertainty in the information used to plan the lift and are not based on dynamic behavior. Thus, regardless of the name applied, these multipliers are not truly impact factors. The use of such load multipliers in lift planning is discussed in Chapter 6.

Horizontal Inertial Loads. Once the load has been lifted, it may be moved in a variety of ways. All of the gantry manufacturers produce side shifting devices that mount on the header beams or cross beams and are used to move the load laterally and all current commercial gantry products are wheel- or roller-mounted, thus permitting longitudinal movement of the system under load. The horizontal loads created by these movements are, of course, inertial in nature. The gantry moves, the load does not, and a gentle swaying begins.

Determination of the longitudinal inertial load due to travel is somewhat subjective. Regardless of the capability of the drive mechanism, the motion is controlled, at least in part, by the gantry operator. Further, as with the lifting and lowering motions, the automated control systems on some gantry models limit the travel acceleration and deceleration rates. To get an order of magnitude of the horizontal loading that may occur due to travel, consider the following example. A gantry leg has a maximum travel speed of about 10 feet per minute (0.167 foot per second; 0.051 meter per second). The gantry system's hydrostatic drive is infinitely variable, so we do not have a defined rate of acceleration when the system is operated prudently. The maximum speed v, the rate of acceleration a, and the time t required to accelerate from rest to the maximum speed are related by the equation $v = at$. If we assume that the time needed to accelerate to the maximum speed is one-half second, then $0.167 = a\ 0.5$, or $a = 0.333$ foot per second per second (0.102 meter per second per second). The acceleration of gravity g is 32.174 feet per second per second (9.807 meters per second per second), so $a = 0.0104\ g$, or 1.04% of the supported load. We can round this value off to 1.0% for practical use in evaluating the loads to which a gantry system may reasonably experience.

A similar analysis can be made in the lateral direction. Here we are concerned with the performance of the side shifter mechanism. As with system travel, the side shifting motion is controlled by the gantry operator and may be limited by the control system. Again, we will consider an example to derive an understanding of the potential lateral dynamic load. Assume that a powered lift link has a maximum travel speed of about 7.5 feet per minute (0.125 foot per second; 0.038 meter per second). The hydrostatic drive is infinitely variable, so again we do not have a defined rate of acceleration. We can again use the equation $v = at$ to relate the maximum side shift speed, time to reach that speed, and the rate of acceleration. If we assume that the time needed to accelerate to the maximum speed is once again one-half second, then $0.125 = a\,0.5$, or $a = 0.25$ foot per second per second (0.076 meter per second per second). From this, we find that the lateral acceleration $a = 0.0078\,g$, or 0.78% of the supported load, which we can round to 0.8% for practical evaluation purposes.

The use of the underhung trolley or strand jacks again may develop loads that vary from those resulting from convention gantry system operations. For the underhung trolley application, CMAA (2010) defines a horizontal dynamic load factor, termed the inertia force from drives (*IFD*), as given in Eqs. 4.8*a* (in USCU) and 4.8*b* (in SI units).

$$IFD = \left(2.50/32.2\right)a = 0.078a \geq 0.025 \qquad (4.8a)$$

$$IFD = \left(2.50/9.807\right)a = 0.255a \geq 0.025 \qquad (4.8b)$$

This factor is a function of the acceleration rate of the trolley a in feet per second per second (Eq. 4.8*a*) or in meters per second per second (Eq. 4.8*b*) and is based on 250% of this acceleration. The specified lower bound value of $IFD = 0.025$ equates to a value of $a = 0.322$ feet per second per second (0.098 meters per second per second). Note that this dynamic load acts in the direction of travel of the trolley. This specification does not define a dynamic load that acts perpendicular to the beam on which the trolley runs.

As with vertical loading, there are currently no industry standards for horizontal dynamic loads due to lifting with strand jacks. In these more sophisticated gantry system applications, it is incumbent upon the lift planner/engineer to evaluate the application and apply suitable dynamic load factors to the design or selection of the equipment.

The discussion of dynamic loads presented in this section is based on the performance capabilities of the most common gantry products currently on the market. The values discussed here may not be characteristic of specially designed gantry equipment or of future models. However, the principles are fundamental and easily can be applied to evaluate the potential loads that may be developed by other types of equipment and other operating limits.

4.1.5   Misaligned System Travel (Racking)

The travel of a gantry system along its track should be synchronized such that all of the legs move in unison. It is possible, however, that the legs may move out of this perfect alignment, whether due to differences in mechanical functions within the drives or operator error. The most common manifestation of such a misalignment is that the legs on one side of the system move ahead of the legs on the other side. This behavior is commonly called racking. (The term skewing is used when referring to this type of behavior in overhead bridge, gantry, or portal cranes.)

Consider the basic four-leg system shown in Fig. 4.14. A typical gantry system consisting of four legs and two header beam is used to lift and travel with a load that is rigged from four padeyes. As shown, the four legs are at the corners of a rectangle and the lift links are directly above the load's padeyes, which are also arranged as a rectangle. The rigging hangs plumb in this static condition. The center of gravity of the load is assumed to be centered in both horizontal directions for this initial discussion. This is the ideal arrangement and the static and dynamic forces that may be imposed on the gantry system are as discussed in the previous sections.

Now consider the condition in which the legs on one side of the gantry system move ahead of the legs on the other side by a distance R, as illustrated in Fig. 4.15. The header beams are no longer perpendicular to the direction of travel and, as a result, the lift links are no longer directly above the padeyes. The effect of racking is illustrated in two steps. Fig. 4.15a shows the misalignment R of the gantry legs and the resulting clockwise rotation of the header beams from the perpendicular. This movement of the header beams results in a longitudinal misalignment of the lift links relative to the padeyes on the load. As a result, the slings go out of plumb, thus creating horizontal forces F_{hx} as illustrated in Fig. 4.16. These four forces F_{hx} induce a clockwise torsion on the suspended load. The load then undergoes a clockwise rotation, illustrated in Fig. 4.15b, thus developing the lateral misalignments and corresponding forces F_{hy}. The final longitudinal and lateral horizontal padeye-to-lift link misalignments are lengths x and y. Equilibrium is reached when the clockwise torsion T_{cw} due to the forces F_{hx} is equal to the counterclockwise torsion T_{ccw} due to the forces F_{hy}.

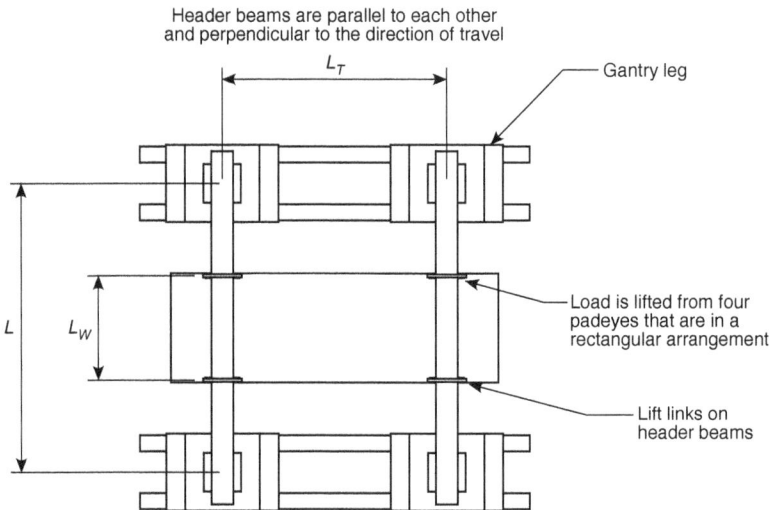

Figure 4.14 Gantry System with Uniform Travel

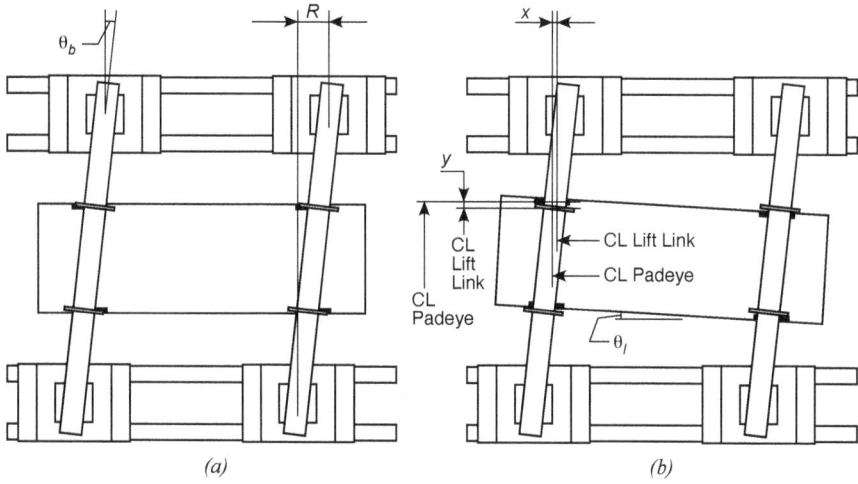

Figure 4.15 Gantry System Subjected to Racking

Using the dimensions shown in Figs. 4.14, 4.15, and 4.16, we can write Eqs. 4.9 through 4.15.

$$\theta_b = \sin^{-1}\left(R/L\right) \tag{4.9}$$

$$x = \left(0.5L_T \cos\theta_l - 0.5L_W \sin\theta_l\right) - \left(0.5L_T - 0.5L_W \sin\theta_b\right) \tag{4.10}$$

$$y = \left(0.5L_T \sin\theta_l + 0.5L_W \cos\theta_l\right) - 0.5L_W \cos\theta_b \tag{4.11}$$

$$F_{hx} = P\frac{x}{\sqrt{H_r^2 - x^2 - y^2}} \tag{4.12}$$

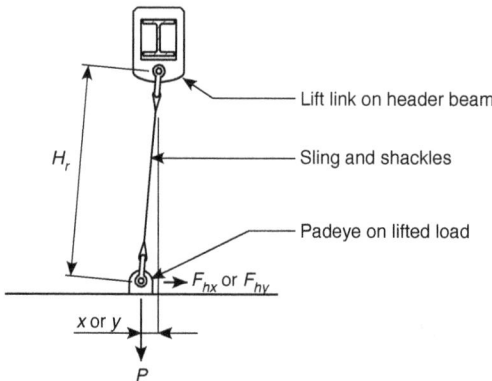

Figure 4.16 Horizontal Force Developed by Rigging Angle

$$F_{hy} = P \frac{y}{\sqrt{H_r^2 - x^2 - y^2}} \tag{4.13}$$

$$T_{cw} = 4F_{hx} \frac{L_W}{2} = 2F_{hx}L_W \tag{4.14}$$

$$T_{ccw} = 4F_{hy} \frac{L_T}{2} = 2F_{hy}L_T \tag{4.15}$$

All of the values needed to calculate the horizontal forces F_{hx} and F_{hy} and the resulting torsions T_{cw} and T_{ccw} are well defined except for the angle of rotation of the suspended load θ_l. The exact value of this angle does not lend itself to direct calculation. However, a reasonable approximate value of θ_l can be calculated using Eq. 4.16.

$$\theta_l = \theta_b \left\{ 1 - \left[0.0298 \left(\frac{L_T}{L_W} \right)^3 - 0.2905 \left(\frac{L_T}{L_W} \right)^2 + 0.9743 \left(\frac{L_T}{L_W} \right) - 0.2152 \right] \right\} \tag{4.16}$$

The solution of numerous verification problems shows that the torsions balance within 2% to 3% using the value of θ_l from Eq. 4.16. The value of θ_l can be refined in the calculations until $T_{cw} = T_{ccw}$ if a more precise solution is required. This process can be accomplished easily by setting up Eqs. 4.9 through 4.15 in a spreadsheet or mathematics application and then manually adjusting θ_l until the two torsional moments are equal.

The significance of the forces developed by racking can be seen when applied to an analysis of the stability of the gantry system. Looking at Fig. 4.15b we can see that both forces F_{hy} at each end of the suspended load act in the same direction. Thus, these racking forces act to destabilize each two-leg gantry in the lateral direction.

The magnitude of the forces developed by racking of the gantry system can best be illustrated by means of an example.

EXAMPLE 4-6
Consider the gantry system shown in Figs. 4.14 through 4.16 with the following layout dimensions and lifted load:

L = 300 inches
L_W = 120 inches, centered on the header beam span
L_T = 180 inches
H_R = 100 inches
Weight of load = 400,000 pounds
Case 1 R = 3 inches
Case 2 R = 15 inches

Case 1 Solution:

Eq. 4.9 gives us θ_b = 0.5730 degree

Eq. 4.16 gives us θ_l = 0.1758 degree, which is refined to 0.1750 degree by solving Eqs. 4.10 through 4.15 and adjusting the angle until $T_{cw} = T_{ccw}$. At this final value of the load rotation angle, Eq. 4.12 gives us F_{hx} = 416 pounds. Given the lift link spread of 120 inches and the header beam span of 300 inches, the header beam longitudinal reaction at each gantry leg is 416 x 120 / 300 = 167 pounds, or 0.17% of the vertical reaction. Eq. 4.13 gives us F_{hy} = 278 pounds, which is 0.28% of the vertical load P. Thus, racking the gantry system by 3 inches, or 1% of the header beam span, develops a longitudinal force on the gantry leg equal to 0.17% of the vertical load and a lateral force equal to 0.28% of the vertical load.

Case 2 Solution:

Eq. 4.9 gives us θ_b = 2.8660 degrees

Eq. 4.16 gives us θ_l = 0.8793 degree, which is refined to 0.8484 degree again by solving Eqs. 4.10 through 4.15 and adjusting the angle until $T_{cw} = T_{ccw}$. At this final value of the load rotation angle, Eq. 4.12 gives us F_{hx} = 2,102 pounds. Given the lift link spread of 120 inches and the header beam span of 300 inches, the header beam longitudinal reaction at each gantry leg is 2,102 x 120 / 300 = 841 pounds, or 0.84% of the vertical reaction. Eq. 4.13 gives us F_{hy} = 1,402 pounds, which is 1.40% of the vertical load P. Thus, racking the gantry system by 15 inches, or 5% of the header beam span, develops a longitudinal force on the gantry leg equal to 0.84% of the vertical load and a lateral force equal to 1.40% of the vertical load.

We can see by Example 4-6 that the lateral forces developed by a relatively small level of racking are likewise relatively small with respect to the other horizontal loads discussed in this section that may act on a gantry system. However, when the racking distance is allowed to become large, the resulting lateral force will become significant, approaching the magnitude of other lateral loads that may act on the gantry system. It is noted that the results of these calculations will vary for problems with different proportions, but the general trend observed here will remain. This comparison highlights the importance of operating the gantry system so as to minimize racking.

This discussion and the accompanying example are based on the simplest arrangement in which the load and gantry system are both symmetrical in both horizontal directions. The performance of a racking analysis of a load and gantry system that is asymmetric is more complex, but follows the same basic principles outlined here. Since a racking analysis is not a part of routine gantry system lift planning, the more complex equations needed to perform such an analysis are not derived here.

4.1.6   Wind and Seismic Loads

Wind loading on a gantry system most often ranges from insignificant to non-existent. When a gantry system is used outdoors, the wind climate to which the system is exposed during a lift is very mild. Heavy lifts are typically forbidden when the wind speed exceeds 20 to 30 miles per hour (32 to 48 km/h), gantries are relatively short [about 48 feet (14.6 meters) tall for the largest machines currently on the market], and the types of items most commonly handled by gantries, such as machinery components, pressure vessels and the like, are very heavy relative to the sail area presented to the wind. When used indoors, of course, the gantry and its payload are completely sheltered from the wind. Because of this limited exposure to wind loading, the development of a specialized calculation method is not necessary. Standard calculation methods for the determination of wind loads may be used when needed. Example 4-7 demonstrates the application of the widely accepted wind load calculation method of ASCE/SEI 7-10 (ASCE 2010).

EXAMPLE 4-7

Calculate the wind loading on a flat-sided object using the method defined in ASCE/SEI 7-10. The lifted load and site conditions are defined as follows.

Height = 10 feet (3.05 meters)

Length = 50 feet (15.24 meters)

Weight = 200 tons (1,780 kN)

Vertical position - centered 20 feet (6.10 meters) above the ground

Site description - clear, open site (Exposure C)

Basic wind speed V = 25 miles per hour (40 km/h)

Since the application of ASCE/SEI 7-10 is not unique to gantry lift engineering, the calculations will not be detailed here. Only the results are shown for brevity.

Design wind pressure, p = 1.96 psf (94 Pa)

Projected area, A_f = 10 x 50 = 500 square feet (46.45 square meters)

Total wind force, F = 1.96 x 500

= 981 pounds (4,366 N)

= 0.245% of the lifted weight

This wind load is seen to be substantially less than the other horizontal loads discussed in the preceding sections. Thus, it is often reasonable to ignore wind loading in the design of a gantry installation.

One obvious exception to this statement is the case of lifting or supporting an object of unusually large area. The vessel shown in Fig. 4.17 presents an unusually large sail area to the wind, thus opening the possibility that the gantry system will be subjected to a significant lateral load from the wind. In this situation, a thorough calculation of wind loading must be made as a part of the lift planning. This may include establishing a maximum wind speed [lower than the 20 to 30 mph (32 to 48 km/h) mentioned previously] above which the lift may not proceed in order to limit the effect of wind loading.

An easy-to-apply approximation can be developed by which we can quickly check the significance of wind load on the item being lifted. Let us assume that, for ordinary lift practices, we will not be concerned with wind if the wind load

Figure 4.17 Gantry Lifting an Unusually Large Vessel *(J&R Engineering Co., Inc.)*

does not exceed 0.25% of the weight of the lifted load. Once we know the weight W of the lifted load and the maximum likely wind speed V that may occur during the lift, we can use Eq. 4.17a or 4.17b to compute the sail area A_w that corresponds to a wind load approximately equal to 0.25% of the weight of the lifted load. As long as the sail area of the item being lifted that is presented to the wind is less than A_w, the wind load for the stated speed may be disregarded under ordinary circumstances.

$$A_w = \left(1,900 V^{-2.08}\right) W \qquad (4.17a)$$

$$A_w = \left(522 V^{-2.08}\right) W \qquad (4.17b)$$

Eq. 4.17a is for use with USCU in which the wind speed V is in miles per hour, the weight W is in tons, and the calculated sail area A_w is in square feet. Eq. 4.17b is for use with SI units in which V is in kilometers per hour, W is in tonnes (1 tonne = 1,000 kg), and A_w is in square meters. One can also enter Eqs. 4.17a and 4.17b with values of the weight W and sail area A_w and solve for the maximum acceptable wind speed V for the lift.

To better visualize this relationship, a plot of the weight multiplier portion of Eq. 4.17a (the quantity in parentheses) is graphed in Fig. 4.18.

Eqs. 4.17a and 4.17b were developed by calculating the wind load on a flat-sided object using the provisions of ASCE (2010). The wind speeds used in this study ranged from 5 to 30 mph (8 to 48 km/h), the lifted load ranged from 20 to 850 tons (18 to 770 tonnes) and the height of the load above grade for each calculation, 25 to 45 feet (8 to 14 meters) was selected to correspond with typical gantry system heights for the weight being lifted. The wind area was calculated by multiplying the weight by a constant to arrive at a wind load that did not exceed 0.25% of the weight of the lifted load. Eqs. 4.17a and 4.17b represent smoothed curves of these constants.

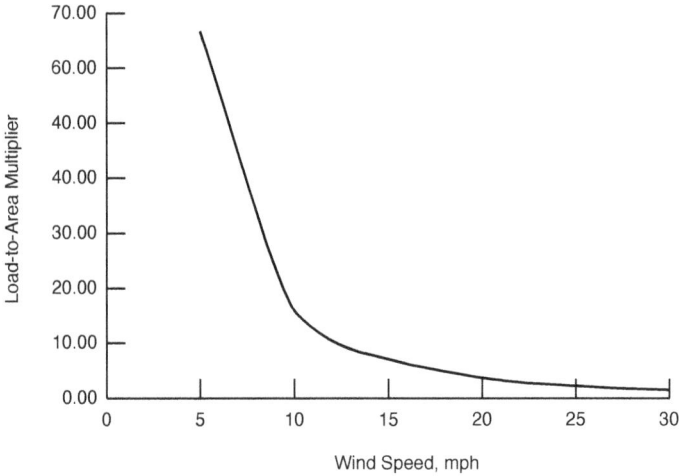

Figure 4.18 Plot of Eq. 4.17*a* Wind Area Multiplier

We can observe the application of this approximation by comparison to the lift examined in Example 4-7.

EXAMPLE **4-8**

Lifted load weight W = 200 tons

Lifted load sail area = 10 x 50 = 500 square feet

Wind speed V = 25 miles per hour

Eq. 4.17*a* is solved to find the area in square feet that corresponds to a wind load equal to approximately 0.25% of the weight:

$$A_w = \left(1,900V^{-2.08}\right)W = \left(1,900 * 25^{-2.08}\right)200$$

$$A_w = 469.97 \text{ square feet}$$

As the actual area of the load is 500 square feet and Example 4-7 resulted in a wind load equal to 0.245% of the lifted weight, we see a good correlation between the detailed calculations following ASCE (2010) and the approximation defined by Eq. 4.17*a*. (Solution of these examples in SI units show a similarly acceptable correlation.)

The need to consider seismic forces in the design of any type of temporarily placed lifting equipment is relatively rare. When one considers the short duration of a lift, the probability of a significant earthquake occurring while the load is in the air is extremely small. In the case of a hydraulic gantry system, the consideration of seismic forces in any region of high seismicity will generally show that the system will topple. The lateral forces developed in a significant seismic event are simply greater than the inherent stability of the gantry legs can resist. Even if the bases of the gantry legs are secured by means of structural anchorages, the strength capacities of the other elements of the legs, such as the lift booms or lift cylinders, will most likely not be able to carry the lateral forces that result from an earthquake.

The consideration of seismic forces is generally unnecessary for normal gantry lift engineering. If a governing authority requires that the lift engineering address seismic forces, a project-specific analysis is required. The number of variables that enter into seismic analysis is too great to allow the offering of a generalized lateral force value.

4.1.7 Rigging Length Effects

The length of the rigging between the header beams and the load will, in some cases, affect the magnitude of the loads imposed upon the gantry system from the load and operational effects discussed in the preceding sections. However, because the rigging length also plays into the stability of the gantry system, the selection of length is not always a clear decision. One must understand all of the ramifications of the rigging length and then make a selection based on a balanced analysis.

The length of the rigging enters into the calculations in three of the load sources discussed in this section. These are cross cornering, misaligned rigging, and racking. In all three cases, an increase in the length of the rigging will reduce the forces imposed on the gantry system from the indicated behavior. Each is discussed briefly here.

The altered load distribution called cross cornering occurs when the four legs of a gantry system do not extend in unison. For a given leg extension difference, the magnitude of the cross cornering effect is a function of the torsional stiffness of the load, the axial stiffness of the rigging, the bending stiffness of the header beams, the span of the header beams, and the positions of the lift links. The axial stiffness of a sling is inversely proportional to its length. That is, if Sling A is twice as long as Sling B and both slings are otherwise identical, the axial stiffness of Sling A is half that of Sling B. As stated in Section 4.1.2, the cross cornering effect is reduced if the axial stiffness of the rigging is reduced, all other aspects of the equipment setup being unchanged. Thus, a greater rigging length is advantageous with respect to diminishing the possible effect of cross cornering.

Misaligned rigging, as discussed in Section 4.1.3, is the condition of the rigging being out of plumb as a result of the lift links on the header beams being eccentric to the lifting attachments on the load. This may occur when initially rigging the load or as a result of drifting the suspended load in order, for example, to achieve an alignment of the load for setting. The out-of-plumb rigging creates a horizontal force equal to $T \tan \theta$, where T is the design rigging tension based on plumb rigging and θ is the angle of the rigging from vertical. We can easily see that for a given horizontal offset of the upper and lower ends of the rigging, the longer the rigging is, the smaller θ will be and, thus the smaller the horizontal force will be.

Thus, again a greater rigging length is advantageous with respect to reducing the horizontal loads that result from out-of-plumb rigging.

The racking of a gantry system due to misalignment of the legs during travel results in the rigging going out of plumb, which creates horizontal forces that are applied to the gantry system in both the lateral and the longitudinal directions. Once the geometry of the racked system and load have been determined, these horizontal forces are calculated with Eqs. 4.12 and 4.13. Noting that the rigging length H_r appears in the denominator of both of these equations, we can immediately see that the horizontal forces due to racking are reduced as the rigging length is increased.

Given the discussion in the three paragraphs above, one might conclude that longer rigging is always advantageous. However, this is one of those situations in which there is no free lunch. While the load effects due to cross cornering, misaligned rigging, and racking may be reduced by increasing the rigging length, two other aspects of the gantry operation are hurt by this change. First, longer rigging will require that the gantry legs be extended to a greater height to make the lift. The greater the gantry leg height, the lower the stability. This reduction of stability occurs because the gantry leg height is the moment arm through which the horizontal forces act to overturn the legs. Details of the calculation of gantry stability are covered in Chapter 5. Second, many gantry models have lower rated loads in the upper stages of extension, so the greater leg extension required to use longer rigging may result in the lift being performed at a lower rated load and, consequently, at a higher percentage of the rated load.

In summary, the issue of rigging length is one of compromise. Changing the length of the rigging may improve the conditions of the lift in one or two areas, but may also diminish the conditions in other areas. Thus we see that the lift planner must understand all of the effects of changes in the rigging length in order to make an optimum choice.

4.2 LOAD COMBINATIONS

Section 4.1 presents explanations of the various loads that may act on a gantry system during the execution of a lift. It is intuitively obvious that all of these loads will not act simultaneously. Further, those loads that do act together at any time during the lift will most likely not all act at their maximum values. This brings us to the subject of load combinations.

Let us first consider vertical loads. The gravity loads will obviously act at all times and at their actual magnitudes. These loads are often well known and their distribution within the gantry system reasonably predictable. In situations where the given weight of the lifted load is suspect, a vertical "impact factor" may be used as a means of compensating for this uncertainty, particularly for the design of header and cross beams. When the additional vertical load represented by this factor is a compensation for uncertainty in the actual gravity load, then this load should also be assumed to act at all times. As discussed in Section 4.1.4, the additional load computed with such a factor is not truly an impact (that is, dynamic) load. On the other hand, if this additional vertical load is truly representative of a dynamic load, such as a load that may occur when downending an item, then this load should be considered to act only in conjunction with the gantry operation that may create the load.

The potential for a variation of the gravity load distribution is introduced by cross cornering, as discussed in Section 4.1.2. Cross cornering can be expected to occur during the lifting or lowering of a load. Since the operator will be alerted to the occurrence of cross cornering by readings on pressure gauges, load indicators or leg extension measurement devices, the unsynchronized leg extensions that cause cross cornering can be corrected as the lifting or lowering proceeds. Thus, an imbalanced load distribution due to cross cornering will most likely be corrected by the operator and will not exist during other operations, such as side shifting or traveling.

True vertical dynamic loads are generally trivial, as discussed in Section 4.1.4. The load factors that are discussed with respect to the use of an underhung hoist are based on compliance with a widely accepted (in the U.S.) industry design specification and are not necessarily indicative of the actual dynamic loading that will occur when lifting with this equipment. The one operation that has the potential of creating true significant vertical dynamic loading is the laying over (or downending) of a load. An abrupt rotational movement of the load, with a resulting "catching" of the load by the gantry, can occur if the lift is not planned and executed so as to maintain full control over the load at all times.

Consider next horizontal loads in the lateral direction. The expected upper bound lateral load due to out of plumb rigging is 1.5% of the vertical load (Section 4.1.3), that due to side shifting motion inertia is 0.8% of the vertical load (Section 4.1.4), and that due to racking at a travel misalignment of 1% of the header beam span may be about 0.3% of the vertical load (Section 4.1.5). The events that generate each of these lateral loads do not occur simultaneously, so these loads are by no means additive. For example, we don't side shift while drifting a suspended load to get it over the anchor bolts.

A similar evaluation can be made of loads in the longitudinal direction. The expected upper bound longitudinal load due to out of plumb rigging is about 1.5% of the vertical load (Section 4.1.3), that due to travel motion inertia is about 1.0% of the vertical load (Section 4.1.4), that due to overcoming the rolling resistance of the leg is about 1.5% of the vertical load (Section 4.1.3), and that due to racking at a travel misalignment of 1% of the header beam span may be about 0.2% of the vertical load (Section 4.1.5). As with the lateral loads, these longitudinal loads do not act simultaneously, so these loads are likewise not additive.

Just as all of the longitudinal loads do not act together and all of the lateral loads do not act together, it is obvious that the maximum longitudinal loads and the maximum lateral loads also do not act simultaneously. A gantry system does not generally travel and side shift at the same time, so the related dynamic forces will occur in one direction or the other, but not both. The rigging can possibly be out-of-plumb in both directions at once during the initial pick, but there is a reduced likelihood that the maximum values will occur in both directions together. Similarly, the supporting surface (usually track beams) can be out of level in both directions, but being out-of-level to the maximum permitted tolerance in both directions is unlikely.

Table 4.1 shows a list of the primary categories of loads that may act on a gantry system and the major types of gantry system operations. Bullets indicate the relationships between gantry system operations and resulting loads. Note that this table does not indicate the loads that may possibly occur in each operation. Rather, the table indicates the more probable load/operation relationships. Out-of-level track is a setup issue and is not related to the operation of the system and, thus, is not included in this table.

For the gantry system user, these load combinations are generally not important considerations for the purpose of lift planning and execution. The gantry user is typically most concerned with the sizing of header and cross beams, track, and the track supports. The design of these elements typically must consider horizontal loads acting in one direction only, so appropriate upper bound values must be used. These design issues are discussed in detail in Chapter 6. Understanding the combinations of horizontal loads acting in both directions is of primary concern to the gantry leg manufacturer. However, it is usually helpful for the user of lifting equipment to have at least a working knowledge of the engineering basis for the design and rating of that equipment.

A practical requirement for the design loading of a gantry leg will consider the maximum dynamic load acting in one direction in combination with some reduced dynamic load in the perpendicular direction. The supporting surface should be taken as out of level to the limit in the same direction as the maximum dynamic load. Two load combinations are required with this approach, one with

TABLE 4.1 Load Occurrence vs. Gantry System Operation

Type or Cause of Load	Lift or Lower	Travel	Side Shift	Upend / Downend	Drift
Gravity	•	•	•	•	•
Cross Cornering	•				
Out-of-Plumb Rigging - Longitudinal	•	•		•	
Out-of-Plumb Rigging - Lateral	•		•		•
Vertical Dynamic				•	
Longitudinal Dynamic		•		•	
Lateral Dynamic			•		
Racking		•			
Environmental	•	•	•	•	•

the maximum dynamic load in the longitudinal direction and the other with the maximum dynamic load in the lateral direction.

Conspicuously absent from this discussion are any recommendations of load factors from which suitable gantry system design load combinations can be computed. The current practice in the design of traditional civil engineering structures, such as buildings and bridges, is based on sets of load combinations defined in standards such as ASCE (2010). Two sets of load combinations are given, one set to be used in conjunction with structural design following the Allowable Strength Design (ASD) methodology and the other for use in conjunction with design following the Load and Resistance Factor Design (LRFD) methodology. More significant to our discussion here, however, is that the load combinations defined in the standards for building and bridge design are based on many decades of practice, extensive research, and detailed statistical analysis. This development of load values, load factors, and load combinations is made possible by the availability of a substantial volume of published data about the characteristics of the loads to which these types of structures may be subjected over their useful lives.

Similarly documented loading data do not exist for telescopic hydraulic gantry systems. As indicated in the discussions in Section 4.1 of the various types of loads to which a gantry system may be subjected, we can make reasonable estimates of the upper bound values of these loads that may occur during normal (or close to normal) gantry system operation based on fundamental physical principles along with experience and judgement. However, these estimates do not give us a statistical picture of the probabilities of various load occurrences that are needed to meaningfully perform the analyses necessary to arrive at load factors and combinations similar to those available for bridge and building design. For the present, then, we simply do not have the same tools at our disposal as do the bridge and building designers.

The development of load combinations must be done in conjunction with the development of structural and mechanical design factors and design methodologies. Collectively, these design factors and methods define the reliability of the gantry leg and other components of the gantry system, which is a fundamental safety issue. In this context, the term "reliability" refers to the probability of survival of a structure or machine over the course of its life. A probability of survival of near 100% is needed for lifting equipment.

When discussing structural and mechanical reliability, the reader must note that a probability of survival of 100% is not possible. In statistical analysis, a survival of, for example, 99.9999% can be achieved, but an absolute 100% cannot. In other words, if there is any possibility at all that an event, such as a failure of any particular component, can occur, then the calculations will recognize that possibility and yield a probability of survival of something less than 100%.

As such and given the lack of hard data, this work of developing load factors, load combinations, and design factors is a responsibility that ultimately falls to either the gantry system manufacturer or a specification- or standard-writing body, such as the American Society of Mechanical Engineers (ASME). In the absence of a governing specification or standard, the manufacturer must assume responsibility for the development of appropriate design criteria. When a design standard applicable to telescopic hydraulic gantry systems is ultimately developed, the resulting load combinations and design factors will be a product of a collaborative effort of gantry system manufacturers, users, and other subject matter experts who can pool their knowledge to create reasonable approximations of loading spectra, design performance requirements, and the like.

Structural reliability has emerged in recent years as an academic subject taught as a part of civil and mechanical engineering programs. For the reader interested in further study of this subject, Nowak and Collins (2013) is an excellent text on the field. The material in this book presumes a background in structural or mechanical engineering, but does not require a formal education in statistics and probability. A discussion of the development of design factors specific to lifting equipment can be found in a paper written by the author (Duerr 2008).

One last comment must be made about this discussion of loads and load combinations. During the normal and proper operation of a gantry system, some of the loads discussed in Section 4.1 will be trivial and others will be relatively small. Further, these loads are likely to occur only independently or in combinations in which their maximum values are not reached. However, if the gantry system is not operated properly, these loads can increase in magnitude and occur in combinations that may create significant hazards. For example, we saw in Example 4-6 that the horizontal forces due to racking can reach unacceptable levels if the racking misalignment becomes unreasonably large. This point highlights the importance of competent lift planning and skilled operation of the gantry system.

4.3   REFERENCES

American Society of Civil Engineers (ASCE) (2010), ASCE/SEI 7-10 *Minimum Design Loads for Buildings and Other Structures*, Reston, VA.

American Society of Mechanical Engineers (ASME) (2009), ASME B30.1-2009 *Jacks, Industrial Rollers, Air Casters, and Hydraulic Gantries*, New York, NY.

Crane Manufacturers Association of America, Inc. (CMAA) (2010), Specification #74 – *Specifications for Top Running & Under Running Single Girder Electric Traveling Cranes Utilizing Under Running Trolley Hoist*, Charlotte, NC.

Duerr, D. (2008), "Design Factors for Fabricated Steel Below-the-Hook Lifting Devices," *Practice Periodical on Structural Design and Construction*, Vol. 13, No. 2, American Society of Civil Engineers, Reston, VA.

Nowak, A.S., and Collins, K.R. (2013), *Reliability of Structures*, 2nd edition, CRC Press, Boca Raton, FL.

Young, W.C., Budynas, R.G., and Sadegh, A.M. (2012), *Roark's Formulas for Stress and Strain*, 8th edition, The McGraw-Hill Companies, Inc., New York, NY.

5   Gantry System Stability

Gantry system legs traditionally have been proportioned to fit in confined spaces inside manufacturing facilities. This practice results in legs that are relatively narrow, particularly in the lateral direction, compared to their heights. Consequently, an important concern in the design and use of a gantry system is stability, i.e., its resistance to toppling when subjected, either unintentionally or due to a planned function, to horizontal loading.

Methods are derived in this chapter by which the stability of an individual gantry leg and of a two-leg gantry system can be calculated. This examination addresses the characteristics of the gantry leg, itself, the relationship of stability to track deflections, what the gantry user should consider in the use of a gantry system, and what can be done if stability appears to be lacking. Whereas some aspects of gantry lift engineering, such as header beam design, are familiar to most structural or mechanical engineers, the concepts of gantry leg stability, while not technically complex, are somewhat unique within the realm of lifting equipment design and use. For this reason, a detailed presentation of the stability analysis is made.

Some of the calculation methods discussed in this chapter are illustrated by means of example problems. The gantry leg dimensions, properties, and other characteristics defined for these examples may be regarded as typical values, but they are not representative of any specific gantry leg product or the product line of any particular gantry manufacturer. These values are for demonstration purposes only and must not be used for actual design or lift planning.

The first section of this chapter examines the calculation of the stability of a single gantry leg. This is followed in Section 5.2 by an expansion to the stability analysis of a two-leg gantry system.

5.1   STABILITY OF A GANTRY LEG

The analysis of the stability of a gantry leg is an effort to quantify the resistance of the leg to toppling under the influence of a horizontal force applied at the top of the leg. This horizontal force may act in either the longitudinal or the lateral direction. For most leg designs, stability is provided by the supported vertical load, the dead weight of the leg, and the length and width of the leg's base. The terminology for these calculations is illustrated in Figs. 5.1, 5.2, and 5.3.

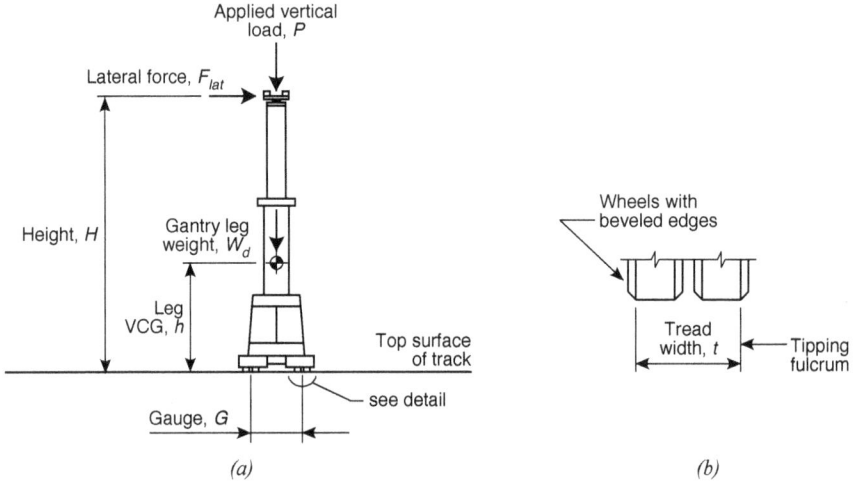

Figure 5.1 Stability Analysis Notation – Lateral Direction

5.1.1   Simplified Stability Analysis

This first analysis will consider the gantry leg in an idealized configuration. The leg is perfectly straight and will not deflect under the influence of a horizontal force acting at the top of the leg. The supporting surface on which the leg bears is perfectly level and will not exhibit differential deflections from one side to the other as the vertical loads at the base of the leg shift from evenly shared side to side to all load on one side as the leg begins to topple (overturn).

Consider first stability in the lateral direction. Toppling of the gantry leg will occur when the overturning effect of the horizontal force acting at the top of the leg exceeds the stabilizing effect of the weight of the leg plus the supported vertical load acting through a moment arm equal to one half of the width of the base. The overturning moment MO_{lat} is equal to the horizontal force acting at the top of the leg multiplied by the height from the top of the track to the top of the leg. The stabilizing moment MS_{lat} is equal to the applied vertical load and the leg's self-weight multiplied by their respective righting arms. For this simplified analysis, the applied vertical load and the gantry leg self-weight both have the same righting arm a, which is one half of the width of the base.

These overturning and stabilizing moments acting in the lateral direction can be written in equation form as follows using the notation illustrated in Fig. 5.1.

$$MO_{lat} = F_{lat}H \tag{5.1}$$

$$a = \frac{G}{2} + \frac{t}{2} \tag{5.2}$$

$$MS_{lat} = a\left(P + W_d\right) \tag{5.3}$$

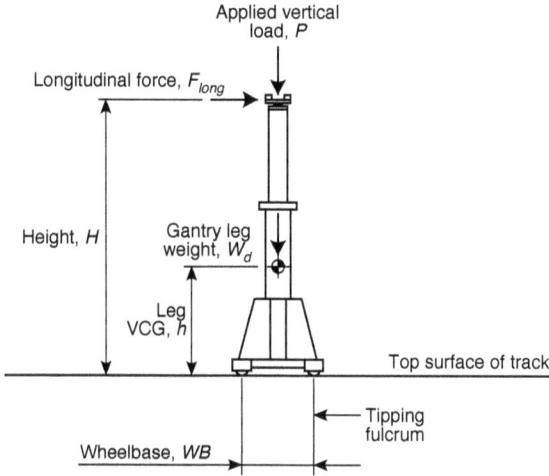

Figure 5.2 Stability Analysis Notation – Longitudinal Direction with Two Axles

These equations can be written for the calculation of the overturning and stabilizing moments in the longitudinal direction using the notation illustrated in Fig. 5.2.

$$MO_{long} = F_{long}H \tag{5.4}$$

$$a = \frac{WB}{2} \tag{5.5}$$

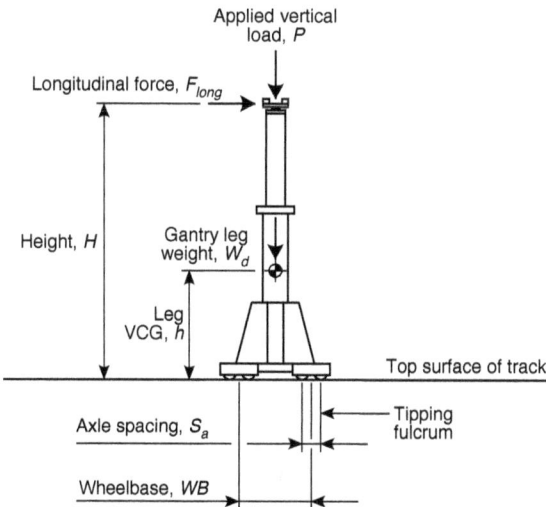

Figure 5.3 Stability Analysis Notation – Longitudinal Direction with Four Axles

$$MS_{long} = a\left(P + W_d\right)$$ (5.6)

The gantry leg shown in Fig. 5.2 has one axle line of wheels at each end of the base. Some gantry leg models utilize wheel assemblies that have two axle lines at each end, as illustrated in Fig. 5.3. The tipping fulcrum here will coincide with the outermost axle line, provided that the wheel assemblies are rigidly connected to the gantry leg base weldment, which is the most common arrangement, and that the affected components are of adequate strength. The only computational change is in the value of the righting arm a. Eq. 5.5 is replaced by Eq. 5.7 for this arrangement.

$$a = \frac{WB}{2} + \frac{S_a}{2}$$ (5.7)

The length of the righting arm a for other gantry leg styles, such as those with a continuous roller chain, rather than wheels, must be determined based on the geometry and mechanical function of the base. If any type of load equalizing attachment is used as a part of the wheel or roller chain assembly, the value of the righting arm a must consider the mechanism effect. For example, if the double-axle wheel assemblies of the gantry leg shown in Fig. 5.3 were attached to the leg base weldment with pivots, rather than rigid connections, then the length a must be based on the center-to-center spacing of the pivots, rather than the distance between the outermost axle lines.

Examination of Eqs. 5.1 through 5.7 shows that this simplified approach to the calculation of the stability of a gantry leg is nothing more than a geometric ratio. The ratio of the leg's height to one half of its base width or length equals the ratio of the sum of the vertical loads to the horizontal force that will initiate toppling of the leg. While this approach provides a quick means of comparing the inherent stability of two similar gantry leg models, a more realistic evaluation of the stability requires consideration of additional characteristics of both the gantry leg and its setup, as described in the next section.

5.1.2   Detailed Stability Analysis

If the gantry leg was perfectly straight and supported on track that was perfectly level and rigid both laterally and longitudinally, then the simplified stability analysis presented in the preceding section would provide a valid solution. This, of course, is not the case. Three effects serve to diminish the stabilizing moment of the gantry leg. These are drift of a telescopic lift boom, deflection of both a lift boom and a lift cylinder, and an out-of-level support surface. Note that the supporting surface may be out of level due to the elevation tolerances permitted for installation or as a result of unequal deflections of the surface.

The effect of lift boom drift is illustrated in Fig. 5.4. A lift boom consists of two or more coaxial sections that slide axially relative to one another. At the maximum extension of the boom, the length of the overlap between each pair of adjacent boom sections is at its minimum. In order to facilitate the telescoping motion, there must be some clearance between sections. These clearances are fixed by the gantry manufacturer, typically through a combination of tight boom fabrication tolerances and the fitting of slider pads during boom assembly. These clearances, while small, allow adjacent boom sections to form a slight angle relative to one another (that is, the boom sections are not perfectly coaxial).

Drift is a product of these clearances between boom sections. The calculation of drift is strictly a geometry problem based on the boom section lengths and the clearances between adjacent boom sections. The calculation procedure is based on the assumption that the boom sections are initially straight and remain straight (boom deflection is addressed separately). Lift boom drift is calculated using the following procedure. The dimensions and notation are as follows and as shown in Fig. 5.5.

C_{TOP} = total clearance between boom sections at the top of the overlap; and,

C_{BOTTOM} = total clearance between boom sections at the bottom of the overlap.

$$C_t = \frac{C_{TOP}}{2} + \frac{C_{BOTTOM}}{2} \qquad (5.8)$$

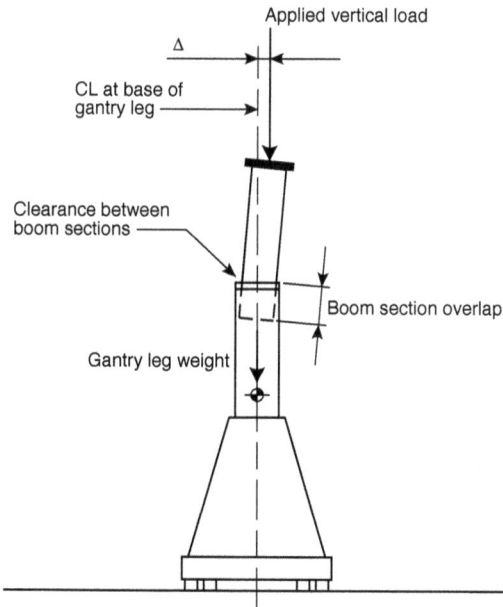

Figure 5.4 Telescopic Lift Boom Drift

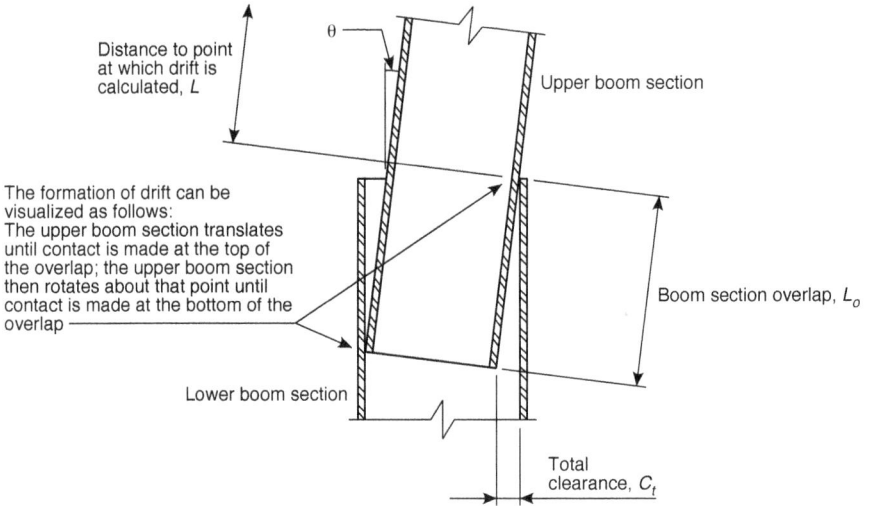

Distance to point at which drift is calculated, L

The formation of drift can be visualized as follows: The upper boom section translates until contact is made at the top of the overlap; the upper boom section then rotates about that point until contact is made at the bottom of the overlap

Lower boom section

Upper boom section

Boom section overlap, L_o

Total clearance, C_t

Figure 5.5 Boom Section Overlap Notation

$$\theta = \sin^{-1}\left(\frac{C_t}{L_o}\right) \tag{5.9}$$

$$\Delta = \frac{C_{TOP}}{2} + L\sin(\theta) \tag{5.10}$$

Consider the following example to illustrate the potential magnitude of drift. Let the boom section overlap L_o equal 24 inches (610 mm), $C_{TOP} = 0.050$ inch (1.27 mm), and $C_{BOTTOM} = 0.050$ inch (1.27 mm). From Eq. 5.8, C_t equals 0.050 inch (1.27 mm). The angle of rotation of one boom section relative to the other is θ and is computed using Eq. 5.9. As shown below, we find a value of θ equal to 0.119° for the given dimensions. (As previously noted, remember that the numbers used in this example, while practical relative to current gantry manufacturing practice, are not specific to any particular gantry leg model or manufacturer.)

$$\theta = \sin^{-1}\left(\frac{0.050}{24}\right) = 0.119°$$

While this rotation angle may seem small, its effect can be large. Define the length of the upper boom section above the overlap as 84 inches (2,134 mm). The drift Δ at the top of the upper section is calculated to be 0.200 inch (5.08 mm) using Eq. 5.10:

$$\Delta = \frac{0.050}{2} + 84\sin(0.119°) = 0.200 \text{ inch}$$

Now consider a second telescoping boom section with the same overlap, clearances, and extension length relative to the first telescoping section. The angle θ of the first telescoping section to the bottom boom section is 0.119° as shown above and, since the overlap dimensions are identical, the angle θ of the second telescoping section to the first telescoping section will likewise be 0.119°. However, the angle of the second section with respect to vertical will be the sum of the angle of the first section to the base section and the angle of the second section to the first section, or 0.239° (the values shown account for round-off). Calculation of the drift Δ at the top of the second section requires that we first compute the drift at the top of the first section (0.200 inch; 5.08 mm) and then compute the additional drift of the second section relative to the first section (0.375 inch; 9.52 mm). The total drift at the top of the second section is equal to the sum of these two values, which is 0.575 inch (14.60 mm).

We can add a third telescoping section, again with the same overlap, clearances, and extension length, and repeat the calculations to find the angle of this third section to be 0.358° from vertical and the drift at the top of this third section to be 1.125 inches (28.57 mm).

We can easily see from this example how boom drift develops for a multiple section lift boom. The three-stage boom considered here is not particularly tall. The length of the base section for such a boom would be about 100 inches (2,540 mm), making the total extended length of the boom to be 352 inches (8,941 mm). The drift using the example dimensions was found to be 1.125 inches, or 0.32% of the total extended boom height. If we consider a gantry leg with a lateral righting arm *a* equal to 24 inches (610 mm), a drift of 1.125 inches equates to a 4.69% reduction in the length of the righting arm for the applied vertical load, which means we have a 4.69% reduction in that portion of the stabilizing moment attributable to the applied vertical load.

The example illustrated with reference to Fig. 5.5 uses the gross overlap between sections in the drift calculations. This is not always the approach used in design practice. Boom sections are often fitted with slider pads, as shown in Fig. 5.6. These pads serve two functions. First, they enable the boom manufacturer to adjust the fit up of the boom sections to minimize the clearances. Second, they define the points of load transfer between boom sections.

Boom strength calculations for the design of telescopic booms used in mobile cranes (e.g., SAE 2007) are based on the assumption that the reactions between boom sections occur at the centers of the bearing lengths of the slider pads, not at the ends of the boom sections. Adoption of such a model for the analysis of a gantry lift boom is reasonable and practical. The various length dimensions, such as the overlap, are thus based on these slider pad bearing length centers. It is appropriate, then, to perform boom drift calculations using length dimensions based on the centers of the slider pad bearing lengths for analytical consistency. This will result in somewhat greater computed drift values as compared to those resulting from calculations based on the gross overlap lengths. Considering the significance of boom drift on gantry leg stability and the absence of published

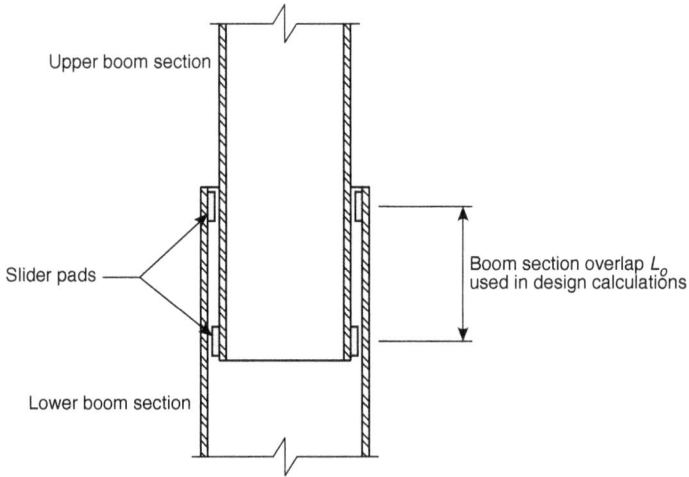

Figure 5.6 Boom Section Overlap Detail

experimental or observational data by which calculation models can be verified, this more conservative approach is warranted.

An alternate slider pad design is illustrated in Fig. 5.7. Here, a single long slider pad mounted to the inside of the lower boom section is used at each location. The interaction of the two boom sections is more like that illustrated in Fig. 5.5 in which the contact points between boom sections are not as clearly defined as when relatively short slider pads are used. Experience shows that this design is an acceptable detail. Note that the overlap length L_o does not change during boom extension with this type of slider pad.

It is noted that there are no current industry standards that define a limit to lift boom drift. The SC&RA *Recommended Practices for Hydraulic Jacking Systems* (SC&RA 1996) suggested a drift limit of 0.375 inch per 10 feet of boom height (9.38 mm per 3,000 mm of height), or about 0.31% of the height, substantially identical to the value found in the example problem. However, this provision was not maintained in SC&RA (2004). ASME B30.1-2009 (ASME 2009) does not specify a drift limit, but does require that drift be considered in gantry leg strength and stability calculations.

Drift due to the clearances between sections is a characteristic of lift booms only. Due to the machining tolerances used in the manufacture of hydraulic cylinders combined with the fit of the seals, drift of a lift cylinder is nominally zero and need not be considered in stability calculations.

It is obvious that drift of a lift boom will create a lateral shift in the center of gravity of the gantry leg. Because the bulk of the displacement of the boom tip occurs in the upper sections and the overall displacement is relatively small, it is a reasonable approximation to ignore this center of gravity shift due to drift.

Although Fig. 5.4 illustrates boom drift in the lateral direction, drift occurs in the longitudinal direction as well.

Upper boom section

Long slider pad mounted — to lower boom section

Boom section overlap L_o used in design calculations

Lower boom section

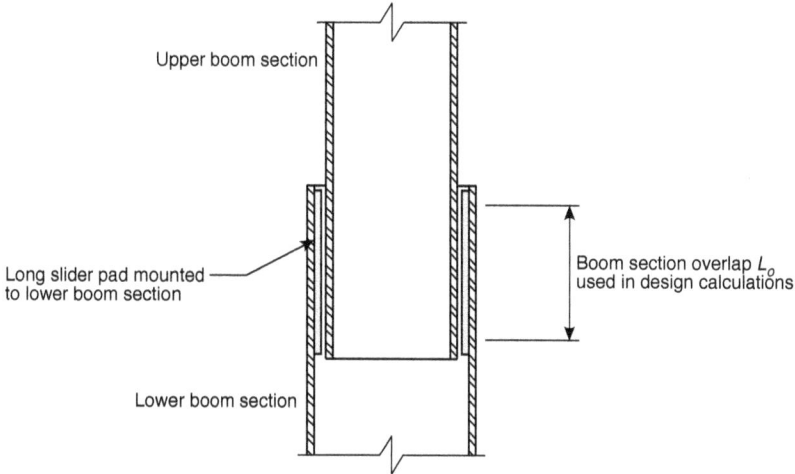

Figure 5.7 Boom Section Overlap with Long Slider Pad Design

In addition to drift, which is a geometry problem, we must also consider the deflection of the lift boom or lift cylinder. Application of a horizontal force to the top of a lift boom or lift cylinder will result in a global bending deflection of the boom or cylinder. Additionally, local deformations of the walls of the lift boom sections will occur where the slider pads bear. The calculation of these deflections is a very complex problem due to the interaction of the horizontal force and the supported vertical load. Calculation of the deflection of the boom or cylinder due to the horizontal force is easily performed by considering the boom or cylinder to be a cantilever subjected to a load at its free end. However, this deflection creates an offset between the vertical load at the top of the boom or cylinder and its base which, in turn, creates an additional bending moment in the boom or cylinder. Thus, the simple deflection due to the horizontal force is amplified by the additional bending due to the offset the vertical load. This increased deflection is referred to as a P-Δ deflection in structural engineering. The analysis methods by which the deflections of a lift boom or lift cylinder are calculated are generally not of concern to the gantry system user and thus are beyond the scope of this book.

The magnitude of the effect of boom or cylinder deflection on the stability of a gantry leg is somewhat more difficult to assess in this discussion because the magnitude of the deflection is a function of the horizontal and vertical forces applied at the top and the cross-sectional properties of the boom or cylinder sections. Some general statements can be made, however, to provide an "order of magnitude" feel to the problem.

The structural deflection of a lift boom due to global beam-type bending tends to be relatively small. In almost all gantry leg designs, the cylinder that telescopes the boom is contained within the boom. Thus, the innermost boom section must be large enough to fit around the cylinder. The spaces required between boom sections to permit fitting of the slider pads drive the cross-sectional dimensions of the remaining boom sections. The result is that a lift boom is significantly larger, and therefore stiffer, than a lift cylinder. Upper bound lateral or longitudinal deflections due to the combined effects of a horizontal force equal to 1.5% of the vertical load and the P-Δ effect may be on the order of 0.5% to 1.0% of the cantilevered length of the boom. For example, this equates to 1.5 to 3.0 inches (38 to 76 mm) for a boom with a cantilevered length of 25 feet (7,620 mm).

The structural deflection of a lift cylinder will be somewhat greater than that of a lift boom due to its more slender cross sections. Using the same horizontal force of 1.5% of the vertical load, lift cylinder deflections may be on the order of 2.0% to 3.0% of its cantilevered height (again including the effect of the P-Δ moment). This equates to 6.0 to 9.0 inches (152 to 229 mm) for the same cantilevered height of 25 feet (7,620 mm).

As was noted with respect to lift boom drift, deflection of a lift boom or lift cylinder will also create a lateral shift in the center of gravity of the gantry leg. It is a reasonable approximation to ignore this center of gravity shift due to deflection, as was discussed with respect to drift. Last, the effect of deflection applies in both the lateral and longitudinal directions.

The third effect to consider is that of an out-of-level support surface. Fig. 5.8 illustrates how the righting arms of both the supported load and the gantry leg weight are decreased when the supporting surface is out of level. This effect can be particularly significant for a gantry leg with a gauge that is narrow relative to its height. Consider, for example, a gantry leg that is 33 inches (838 mm) wide out-to-out across the wheels and is extended to a height of 25 feet (300 inches; 7,620

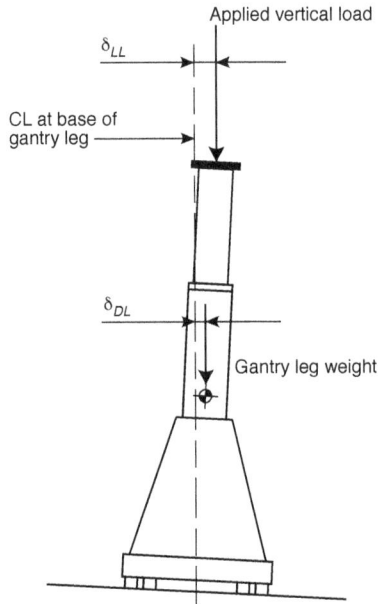

Figure 5.8 Out-of-Level Support Surface

mm). If the supporting surface is 0.125 inch (3.18 mm) lower on one side relative to the other, the lateral shift of the top of the leg δ_{LL} is computed as follows.

$$\delta_{LL} = 300 \text{ x } 0.125 / 33 = 1.14 \text{ inches } (7,620 \text{ x } 3.18 / 838 = 28.9 \text{ mm})$$

If we neglect drift and deflection of the leg for this example, the righting arm a of the gantry leg on a level surface is half of its width, or 16.50 inches (419 mm). Allowing the supporting surface to be out of level by 0.125 inch (3.18 mm) reduces the righting arm to $16.50 - 1.14 = 15.36$ inches (390.1 mm), a reduction of 6.9%, which produces a corresponding reduction in the stabilizing moment. And as with drift and deflection, this effect occurs in both the lateral and longitudinal directions.

The supporting surface for a gantry system is most commonly a set of track beams. If the track beams themselves are supported on a very rigid surface, then the cause of the out of level condition is limited to the tolerances employed when setting up the system. When track beams must span a significant distance, for example, from support stands to a foundation, then deflection of the track beams may contribute to the magnitude of an out of level surface. Methods that can be used to evaluate and minimize the effect of track beam deflection contributing to an out-of-level gantry leg are discussed in Chapter 6.

With these concepts established, we can now write a set of equations by which the lateral and longitudinal stability of a single gantry leg can be calculated. The model to be used is illustrated in Fig. 5.9. Fig. 5.9a shows a gantry leg exhibiting drift and deflection supported on a level surface. Fig. 5.9b takes that leg and rotates it through an angle ϕ due to an out-of-level support.

Let the following symbols represent the forces and dimensions used in the stability analysis.

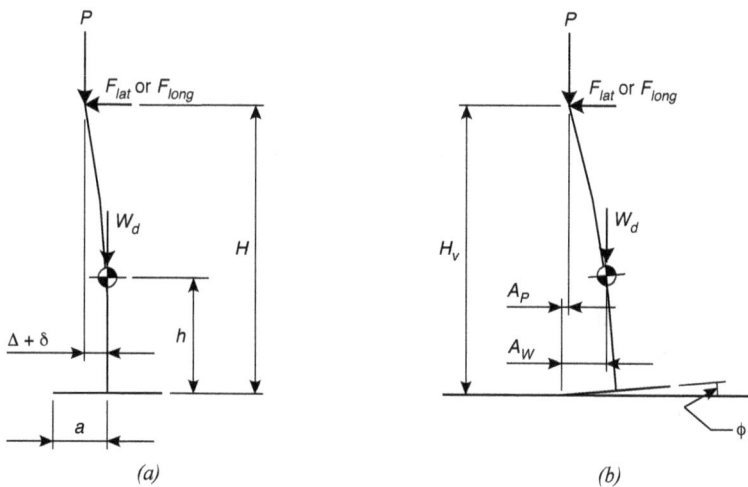

(a)   *(b)*

Figure 5.9 Gantry Leg Stability Analysis Model

P = applied vertical load;
W_d = gantry leg weight;
F_{lat} = lateral force;
F_{long} = longitudinal force;
H = height to top of gantry leg;
H_v = vertical projection of the gantry leg height above the lower side or end, as applicable;
h = gantry leg vertical center of gravity;
G = track beam spacing (gauge);
t = wheel tread width;
WB = wheelbase;
S_a = axle spacing;
E = difference in elevation of track beams across the gauge G;
S = difference in elevation of track beams over the wheelbase WB;
ϕ = angle of support surface from horizontal;
Δ = displacement of header plate due to boom clearances (drift);
δ = displacement of header plate due to boom or cylinder deflection;
A_P = righting arm to applied load P;
A_W = righting arm to gantry leg weight W_d;
MO_{lat} = lateral overturning moment;
MO_{long} = longitudinal overturning moment;
MS_{lat} = lateral stabilizing moment; and,
MS_{long} = longitudinal stabilizing moment.

Any consistent set of units may be used; e.g., all loads in pounds and all lengths in inches, etc.

All of the quantities needed to perform the stability calculations are readily available from the gantry manufacturer or can be determined by the user, with one exception: the gantry leg vertical center of gravity. This is not a fixed value, but rather one that changes with the magnitude of leg extension. In the absence of a vertical center of gravity value from the manufacturer or other reliable source for the leg extension under consideration, the vertical center of gravity of the gantry leg may be reasonably approximated as 30% of the overall height of the leg. Recognizing that the stability calculation results are not particularly sensitive to this value, the use of such an approximation will produce an acceptably reliable result. The values of boom drift Δ and boom or cylinder deflection δ are not necessarily the same in the lateral and longitudinal directions. Care must be taken to use the correct values in these analyses. Last, the calculation of the displacement δ must account for the loading condition created by the out of level base.

Eqs. 5.11 through 5.17 are used to analyze the stability of a gantry leg in the lateral direction.

$$a = \frac{G}{2} + \frac{t}{2} \tag{5.11}$$

$$\phi = \sin^{-1}\left(\frac{E}{G}\right) \tag{5.12}$$

$$A_P = a\cos\phi - H\sin\phi - (\Delta + \delta)\cos\phi \tag{5.13}$$

$$A_W = a\cos\phi - h\sin\phi \tag{5.14}$$

$$H_v = a\sin\phi + H\cos\phi - (\Delta + \delta)\sin\phi \tag{5.15}$$

$$MS_{lat} = A_P P + A_W W_d \tag{5.16}$$

$$MO_{lat} = H_v F_{lat} \tag{5.17}$$

Toppling of the gantry leg is imminent when the overturning moment MO_{lat} is equal to the stabilizing moment MS_{lat}.

As previously noted, most of the quantities used in these calculations are hard values that can be reliably determined by the manufacturer. The only two function-dependent variables are the applied vertical load P and the lateral force F_{lat}. If we define F_{lat} as some percentage of P, as is commonly done for the design of lifting equipment, then the only variable to be defined is this percentage. For reference, ASME (2009) prescribes a design horizontal load of 1.50% of the vertical load.

Similar expressions can be written for the stability of the gantry leg in the longitudinal direction. The value a for the longitudinal direction is defined by Eq. 5.18, rather than Eq. 5.11, adjusted if necessary for pivoting wheel assemblies as discussed on page 114, and the value for ϕ is defined by Eq. 5.19, rather than Eq. 5.12. Eqs. 5.13 through 5.15 can be applied as written to the longitudinal case. The solution for the longitudinal case is then expressed by Eqs. 5.20 and 5.21. Toppling of the gantry leg is imminent when the overturing moment MO_{long} is equal to the stabilizing moment MS_{long}.

$$a = \frac{WB}{2} + \frac{S_a}{2} \tag{5.18}$$

$$\phi = \sin^{-1}\left(\frac{S}{WB}\right) \tag{5.19}$$

$$MS_{long} = A_P P + A_W W_d \tag{5.20}$$

$$MO_{long} = H_v F_{long} \tag{5.21}$$

Once the overturning and stabilizing moments have been calculated, the results can be expressed in a number of different ways. The method preferred by the author is to relate the two moments as a Demand / Capacity (*DC*) ratio, as expressed in Eq. 5.22 for the lateral direction and in Eq. 5.23 for the longitudinal direction. A *DC* ratio of 1.00 indicates that toppling is imminent, so a margin of safety must be incorporated into Eqs. 5.22 and 5.23 for practical application. The use of a Demand / Capacity ratio, rather than a simple "pass - fail" result, allows quantitative comparison among different gantry models (for the user) or evaluation of design changes (for the manufacturer).

$$DC = \frac{MO_{lat}}{MS_{lat}} \leq 1.00 \qquad (5.22)$$

$$DC = \frac{MO_{long}}{MS_{long}} \leq 1.00 \qquad (5.23)$$

A second method of quantifying the stability of a gantry leg is expressed for the lateral direction in Eqs. 5.24 and 5.25. In this approach, the calculated stabilizing moment is used to compute the horizontal force F_{lat} that will initiate overturning. This value is divided by the applied vertical load to express the overturning force as a percent of the applied vertical load. While this method also allows comparisons to be made among different gantry leg models, it does not relate stability to anticipated or probable horizontal loading. Thus, this approach does not have the same practical engineering value as does the approach expressed as Eqs. 5.22 and 5.23.

$$F_{lat} = \frac{MS_{lat}}{H_v} \qquad (5.24)$$

$$\% = \frac{F_{lat}}{P}\left(100\%\right) \qquad (5.25)$$

There is usually very little that can be done by the user of a gantry system to improve the stability of a gantry leg. The deflection and boom drift characteristics of a gantry leg are both functions of the leg's design and manufacture and cannot reasonably be altered by the user. All of the current gantry products have fixed wheelbases, so the longitudinal stability for a given model at a particular height cannot be easily modified. Some models have extendible wheel assemblies (wheel boxes) that can be adjusted to widen the track gauge to improve the lateral stability to some extent.

There are, however, three areas that affect stability that the gantry system user can always control. First, the track layout must be designed to provide firm, uniform support. Second, the contractor can institute improved field quality control procedures to minimize the installed out-of-level of the track beams. Third, the gantry operation can be planned and executed to minimize the generation of horizontal forces acting on the system.

5.1.3   Comparison of the Simplified and Detailed Stability Analyses

The simplified stability analysis relates an overturing moment due to an applied horizontal force to the stabilizing moment of an idealized gantry leg (straight, plumb, and infinitely rigid). The detailed stability analysis introduces modifications to the calculation of the stabilizing moment that account for the effects of boom drift, boom or cylinder deflection, and an out-of-level base. A comparison of the results of the two methods is made in this section to illustrate the significance of accounting for the true geometry of the gantry leg.

The simplified analysis method is little more than a geometric analysis of the leg. The only variable beyond the width-to-height ratio that affects the result is the ratio of gantry leg dead weight to applied vertical load. Eqs. 5.1 and 5.3 can be rewritten with all loads expressed as functions of the vertical load P, where R_h is the ratio of the horizontal force to the vertical load and R_w is the ratio of the dead weight to the vertical load (Eqs. 5.26 and 5.27). The overturning and stabilizing moments are equated in Eq. 5.28 and the ratio of half-width to height, a / H, for the point at which toppling is imminent is then shown as Eq. 5.29 in terms of R_h and R_w. Alternately, we can write Eq. 5.30 to express the value of R_h at which toppling is imminent in terms of a / H and R_w.

$$MO_{lat} = \left(R_h P \right) H \tag{5.26}$$

$$MS_{lat} = a \left(1 + R_w \right) P \tag{5.27}$$

$$\left(R_h P \right) H = a \left(1 + R_w \right) P \tag{5.28}$$

$$\frac{a}{H} = \frac{R_h}{1 + R_w} \tag{5.29}$$

$$R_h = \frac{a}{H} \left(1 + R_w \right) \tag{5.30}$$

An examination of gantry specifications from the various manufacturers shows that the dead weight of the leg varies from 4% to 11% of the rated load of the leg at maximum extension. That is, the value of R_w ranges from 0.04 to 0.11. The gantry specifications also show that the a / H ratio ranges from 0.05 to 0.11 when the leg is fully extended.

Application of the detailed stability analysis requires that we know or can reasonably estimate values for lift boom drift, lift boom or lift cylinder deflection, and the angle from horizontal of the supporting surface. Drift and deflection values are not readily available in the published gantry specifications. As discussed in the preceding section, however, there are "typical" values with which we can work. For the purpose of these examples, we will take the drift of a boom leg as $\Delta = 0.003H$, the structural deflection of a boom leg as $\delta = 0.5R_h H$, and the structural deflection of a bare cylinder leg as $\delta = 1.5R_hH$. Note that the analyses that underlie these relationships are iterative since δ is a function of R_h. Further, although the structural deflections of lift booms and lift cylinders are most logically related to their cantilevered lengths, using the overall gantry leg height H for these estimates is convenient, conservative, and, recognizing the approximate nature of these expressions, does not introduce an unacceptable level of inaccuracy.

(Note that when evaluating the stability of a gantry leg design for its performance under the effect of a specified or assumed horizontal load and the lift boom or lift cylinder deflection behavior is not available, the deflection can be estimated using the relationships discussed above, where the value of R_h is simply that horizontal load divided by the applied vertical load. Considering the lateral direction as an example, $R_h = F_{lat} / P$. If $F_{lat} = 1.50\%$ of P, then $R_h = 0.015$.)

There are no studies available that examine the accuracy with which gantry track systems are leveled in practice. ASME (2009) and SC&RA (2004) both define a limit for the maximum permitted out-of-level of gantry track in the lateral direction as 0.125 inch (3 mm) across the gauge of the gantry leg and as 0.125 inch (3 mm) in 10 feet (3 meters) in the longitudinal direction. These values can reasonably be used to define the upper bound out-of-level of track for stability calculations.

The last value we must establish is the vertical center of gravity of the gantry leg. We will let the height of the center of gravity of the leg $h = 0.3H$, as discussed on page 122.

With these values and relationships established, we can now perform the stability analyses. To facilitate comparison, the results will be expressed in terms of the value of R_h at which toppling is imminent. All input and results are assembled in Table 5.1. The analyses of "Gantry Leg A" use values that may be considered typical for a bare cylinder gantry, the analyses of "Gantry Leg B" use values applicable to a telescopic boom gantry of proportions similar to those of "Gantry Leg A," and the analyses of "Gantry Leg C" use values applicable to a markedly taller telescopic boom gantry.

The results illustrated in these three examples are as expected. The simplified stability analysis, solely based on geometry, gives similar results for the bare cylinder gantry leg and telescopic boom gantry leg of comparable proportions. However, the results diverge when the detailed stability analysis is applied due to the greater leg deflection δ exhibited by the lift cylinder relative to that of the lift boom. In both cases, we can see the magnitude by which the simplified analysis overestimates the stabilizing moment.

The taller telescopic boom gantry (Gantry Leg C) is shown by the simplified analysis to be significantly less stable than the bare cylinder gantry (Gantry Leg A) due to its greater height but similar width. However, the detailed analysis shows that the difference in stability between "Gantry Leg A" and "Gantry Leg C" is trivial when leg stiffness is considered.

The conclusions that can be drawn from these examples are clear. The simplified stability analysis does not produce accurate results due to the lack of

TABLE 5.1 Stability Analysis Comparison

Quantity	Gantry Leg A	Gantry Leg B	Gantry Leg C
Height H, inches	336.00	360.00	480.00
Track gauge G, inches	36.00	36.00	36.00
Tread width t, inches	10.00	10.00	12.00
Gantry leg weight W_d, pounds	5,000	13,500	23,000
Applied vertical load P, pounds	100,000	160,000	225,000
a (Eq. 5.2), inches	23.00	23.00	24.00
MS_{lat} (Eq. 5.3), pound-inches	2,415,000	3,990,500	5,952,000
F_{lat} (Eq. 5.24 using $H_v = H$), pounds	7,188	11,085	12,400
Simplified Analysis R_h	7.2%	6.9%	5.5%
Gantry leg VCG h, inches	100.80	108.00	144.00
Track elevation difference E, inches	0.125	0.125	0.125
Boom drift Δ, inches	0.00	1.08	1.44
Boom or cylinder deflection δ, inches	13.78	7.53	7.76
a (Eq. 5.11), inches	23.00	23.00	24.00
ϕ (Eq. 5.12), degree	0.20	0.20	0.20
A_p (Eq. 5.13), inches	8.05	13.14	13.13
A_w (Eq. 5.14), inches	22.65	22.62	23.50
H_v (Eq. 5.15), inches	336.03	360.05	480.05
MS_{lat} (Eq. 5.16), pound-inches	918,671	2,408,523	3,494,415
F_{lat} (Eq. 5.24), pounds	2,734	6,689	7,279
Detailed Analysis R_h	2.7%	4.2%	3.2%
MS_{lat} (Eq. 5.16) / MS_{lat} (Eq. 5.3)	38.0%	60.4%	58.7%

consideration of boom drift, structural deflections, and out-of-level track. Of particular significance are the inconsistencies when comparing telescopic boom gantry models to bare cylinder models. The best results are obtained by using the detailed stability analysis. In doing so, drift and deflection values that are specific to the gantry leg under consideration must be used.

5.1.4   Stability Calculation by the Energy Method

The stability of a gantry leg may also be computed through application of one of the fundamental natural principles: Conservation of Energy. The overturning of a gantry leg involves the actions of three forces. These are the vertical load supported by the gantry leg, the dead weight of the leg, and the horizontal force acting to overturn the leg, which acts at the top of the leg. The points of application of each of these three forces move as the leg rotates from its initial position to the point at which toppling is imminent. Thus, we can compute the amount of work done (force times distance) by each force.

Consider the gantry leg model illustrated in Fig. 5.10. In the initial position, the supported load P and the leg dead weight W_d are acting vertically downward and a lateral force F_{lat} is acting horizontally to the left. Prior to the beginning of the rotation of the leg, the boom or cylinder will undergo a structural deflection and the boom will develop its drift. We may reasonably take this deflected shape as a rigid body that rotates from its initial position (Fig. 5.10a) through some angle until it is at the point of toppling (Fig. 5.10b).

As the leg rotates from the position of Fig. 5.10a to that of Fig. 5.10b, the vertical forces P and W_d will rise through the distances Δ_v and δ_v respectively and the horizontal force F_{lat} will move through the distance Δ_h. Thus, work is performed by each force and the principle of conservation of energy tells us that the sum of this work must equal zero. The solution from this point forward is simple geometry.

Let the following symbols represent the forces and dimensions used in this energy method stability analysis. This discussion will derive only the equations for calculation of stability in the lateral direction analysis; as with the previous method, the equations for the longitudinal direction are of the same form. Only the input dimensions are changed (e.g., the wheelbase is substituted for the track gauge).

P = applied vertical load;
W_d = gantry leg weight;
F_{lat} = lateral force;
H = height to top of gantry leg;
h = gantry leg vertical center of gravity;

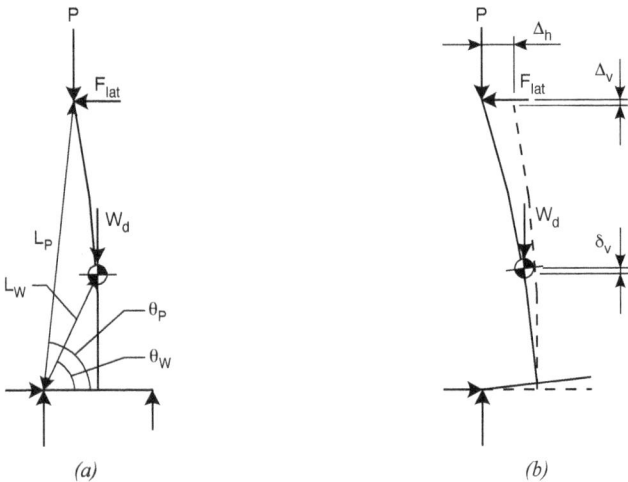

(a)   *(b)*

Figure 5.10 Energy Method Stability Analysis Model

L_P = distance from the tipping fulcrum to load P;
L_W = distance from the tipping fulcrum to the leg center of gravity;
θ_P = angle of line L_P from horizontal;
θ_W = angle of line L_W from horizontal;
G = track beam spacing (gauge);
t = wheel tread width;
E = difference in elevation of track beams across the gauge G;
ϕ = angle of support surface from horizontal;
Δ = displacement of header plate due to boom clearances (drift);
δ = displacement of header plate due to boom or cylinder deflection;
Θ = angle of leg rotation from initial position to toppling;
Δ_h = horizontal movement of header plate as the leg rotates;
Δ_v = vertical movement of header plate as the leg rotates;
δ_v = vertical movement of the center of gravity as the leg rotates;
W_P = work done by movement of load P;
W_W = work done by movement of the weight W_d; and,
W_F = work done by the force F_{lat}.

Using the above notation and with reference to Fig. 5.10, we can write the following equations.

$$L_P = \sqrt{H^2 + \left(\frac{G+t}{2} - \Delta - \delta\right)^2} \qquad (5.31)$$

$$L_W = \sqrt{h^2 + \left(\frac{G+t}{2}\right)^2} \qquad (5.32)$$

$$\theta_P = \tan^{-1}\frac{H}{(G+t)/2 - \Delta - \delta} \qquad (5.33)$$

$$\theta_W = \tan^{-1}\frac{h}{(G+t)/2} \qquad (5.34)$$

$$\Delta_v = L_P\left[\sin(\phi + \theta_P + \Theta) - \sin(\phi + \theta_P)\right] \qquad (5.35)$$

$$W_P = P\Delta_v \qquad (5.36)$$

$$\delta_v = L_W\left[\sin(\phi + \theta_W + \Theta) - \sin(\phi + \theta_W)\right] \qquad (5.37)$$

$$W_W = W_d\delta_v \qquad (5.38)$$

$$\Delta_h = L_P\left[\cos(\phi + \theta_P) - \cos(\phi + \theta_P + \Theta)\right] \qquad (5.39)$$

$$W_F = \frac{F_{lat}\Delta_h}{2} \qquad (5.40)$$

Toppling of the gantry leg will occur when the work done by the horizontal force F_{lat} moving through the distance Δ_h exceeds the work done by the vertical forces P and W_d moving through the distances Δ_v and δ_v, respectively. That is, toppling will occur when $W_F > W_P + W_W$.

This problem cannot be solved directly. The structural deflection of the boom or cylinder δ is a function of F_{lat} and the angle of leg rotation Θ at which $W_P + W_W$ is a maximum that must be determined iteratively. A practical solution method can easily be set up using a mathematics or spreadsheet application. Eqs. 5.31 through 5.38 are solved for a range of values of Θ to determine the maximum value of $W_P + W_W$ that can be reached as the leg rotates. The value of F_{lat} that performs this amount of work can then be calculated using Eqs. 5.39 and 5.41.

$$F_{lat} = \frac{2W_F}{\Delta_h} = \frac{2\left(W_P + W_W\right)}{\Delta_h} \qquad (5.41)$$

A comparison of the detailed analysis method outlined in Section 5.1.2 and the energy method shows that the two methods produce virtually identical results. The detailed analysis method is generally more practical for routine use, since it offers a closed solution.

5.2   STABILITY OF A GANTRY SYSTEM

Section 5.1 examines the stability of a single free-standing gantry leg. In practice, gantry legs are always used in groups of two or more tied together with one or more header beams. This section examines how the components of a functional gantry system interact with respect to stability.

Consider first stability in the lateral direction. A basic two-leg gantry system is illustrated in Fig. 5.11. Of particular importance are the connections between the gantry legs and the header beam. The header plate on most gantry legs in use today is mounted to the leg by means of an articulating connection, typically a pinned connection, a spherical bearing, or a rocker. These styles of header plate mounting create articulating joints between the gantry legs and the header beam. Thus, the two-leg gantry system behaves as a mechanism when toppling occurs in the lateral direction (Fig. 5.12).

The overturning moment MO_{lat} acting on the gantry system of Fig. 5.12 is as given by Eq. 5.17. That is, the overturning moment is equal to the lateral force acting on the system multiplied by the height of that force above the supporting surface of the gantry legs. The stabilizing moment MS_{lat} provided by the gantry system of Fig. 5.12 is simply the sum of the stabilizing moments of each leg, as given by Eq. 5.16. These statements are based on the following assumptions:

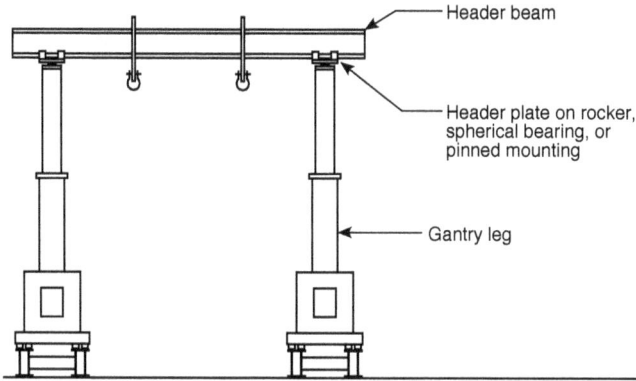

Figure 5.11 Gantry Legs and Header Beam

- The lateral force F_{lat} is small relative to the beam-column strength of the header beam. That is, the compression load induced into the header beam by the lateral force will not cause the header beam to fail structurally prior to the gantry system toppling.
- The connections between the header beam and the gantry leg header plates, either through structural fasteners or by friction, can transmit the horizontal shear reactions due to the lateral force from the header beam to both gantry legs.
- The supporting surfaces of both gantry legs are at the same elevation.

In all but the most unusual arrangements, both gantry legs will be the same model. Therefore, both legs will nominally exhibit the same drift (in the case of boom-type legs) and the same deflection due to the horizontal force. The one

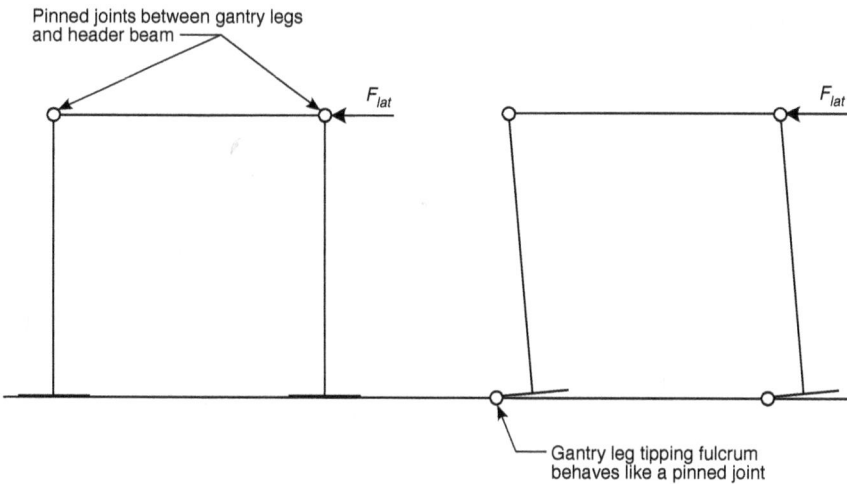

Figure 5.12 Two-Leg Gantry System Modeled as a Mechanism

stability-related value that may differ from one leg to the other is the dimension E by which the track is out of level. The minimum calculated lateral stability occurs when both sides are out of level in the same direction and by the same amount. For planning purposes, the engineer may reasonably assume that the magnitude of E is equal to the maximum value permitted by the governing standards or project requirements.

The difference in the lateral stability of a two-leg gantry system will generally not be altered significantly by the magnitude of the track cross slope as defined by the dimension E. Consider a system composed of two gantry legs with dimensions and properties as listed in Table 5.1 for Gantry Leg B. Performance of the detailed stability analysis gives us the results shown in Table 5.2.

The baseline case is that shown in the first column, where the tracks are dead level. The stability of the system is such that a lateral force equal to 4.4% of the total supported vertical load will bring the system to the brink of toppling. If both tracks are out of level by 0.125 inch (3.18 mm) such that the legs are tilted in the direction of the acting lateral force, the magnitude of that lateral force required to produce instability drops to 4.2%. That is, the track cross slope diminishes the lateral stability by about 5%. Last, if the two tracks are out of level by the same magnitude, but in opposing directions, the reduced stability of one leg is balanced by the increased stability of the other, thus producing the same stability as the system exhibits on level track. All other out of level conditions fall between these two extremes.

This "sum of the parts" stability calculation for a gantry system will not always be applicable. For example, if one track is at a different elevation than the other, the individual leg stabilizing moments cannot simply be added. It is in cases like this that the energy method described in the previous section is the better analytical tool when a stability analysis is deemed necessary.

One gantry manufacturer produces a device designed to enhance the lateral stability of a two-leg gantry system (Fig. 5.13). The device is a spring-loaded

TABLE 5.2 Effect of Track Cross Slope on Lateral Stability

Quantity	Level Track	Identical E	Opposed E
Height H, inches	360.00	360.00	360.00
Track gauge G, inches	36.00	36.00	36.00
Tread width t, inches	10.00	10.00	10.00
Gantry leg weight W_d, pounds each	13,500	13,500	13,500
Total applied vertical load $2P$, pounds	320,000	320,000	320,000
Track elevation difference, E, inches	0.000	0.125	+ or - 0.125
Boom drift, Δ, inches	1.08	1.08	1.08
Boom deflection, δ, inches	7.95	7.53	7.95
Detailed Analysis R_h	4.4%	4.2%	4.4%

Figure 5.13 Gantry Stabilizer Device *(J&R Engineering Co., Inc.)*

strut that is attached to the bottom flange of the header beam with clamps and to the gantry leg with a bracket. The strut changes the behavior of the gantry leg-to-header beam connection from that of a pinned joint to that of a partially restrained joint. A structural component is thus added to the lateral stability of the gantry system and the lateral behavior is that of a frame, rather than the mechanism shown in Fig. 5.12.

The use of this device is limited to situations where the innermost sections of the gantry legs can be extended prior to performing the lifting operation in order to allow the installation of the struts. As such, this device cannot be used on a bare cylinder gantry system.

Due to the relative flexibility of the connections between the gantry legs and the header beam (in the most common applications), the legs can topple in the longitudinal direction more or less independently. Therefore, the longitudinal stability of a two-leg gantry system is evaluated for each leg independently using Eqs. 5.20 and 5.21 and no special considerations come into play. The lower bound longitudinal stabilizing moment occurs when the track slope as quantified by the dimension S is the equal to the maximum value permitted by the governing standards or project requirements.

The evaluation of the longitudinal stability of a four-leg gantry system can become more complex when a four-beam arrangement (Fig. 5.14) is used. In this application, the header beams tie pairs of legs together in the longitudinal direction. The longitudinal stability on one side of the system is now equal to, at a minimum, the sum of the stabilizing moments of the two legs on that side. In the most common applications of four-beam header arrangements, the beams are not rigidly connected to one another, so performing independent evaluations of the longitudinal stability of each side of the gantry system is appropriate. Note that, as with the lateral stability discussion above, the track slope S is not necessarily the same for both legs. The value of S for each leg may be different, perhaps even in opposite directions, when the two legs are located on either side of a joint between

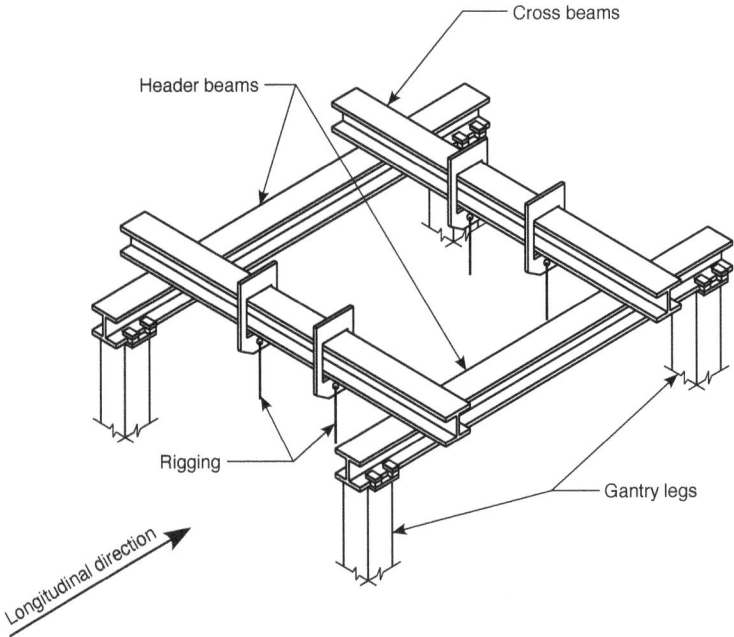

Figure 5.14 Four-Beam Header Arrangement

two track sections. Again, the upper bound value of S should be used for planning purposes.

Additional stability to the gantry system may be developed by the beam system if the connections between the header beams and cross beams are structurally rigid (i.e., bolted or welded). Both lateral and longitudinal stability of the system may be enhanced by the tie provided by the cross beams. Consider, for example, the beam arrangement shown in plan view in Fig. 5.15. Any movement of the top of one gantry leg will be restrained by the rigidity of the beam system and the inherent stability of the other three legs. Put another way, any destabilizing effect that acts on the gantry system will be accompanied by a rigid body motion of the header and cross beams. This, in turn, will mobilize the lateral and longitudinal stabilizing moments of all four gantry legs.

This characteristic of the structurally rigid header and cross beam arrangement is a double-edged sword. On the plus side, such a beam system will, in most cases, increase the natural stability of the gantry system. On the minus side, the analytical evaluation of the gantry system stability becomes a significantly more complex problem. Fortunately, though, calculation of the stability of a gantry system is not routinely required as a part of lift planning and engineering.

The four-beam arrangement may have another effect on stability, depending on the design of the gantry leg's header plate. Some gantry models have header plates that are mounted on rockers and, therefore, articulate in only one direction. Consider, then, the arrangement shown in Fig. 5.16. Here we have the header

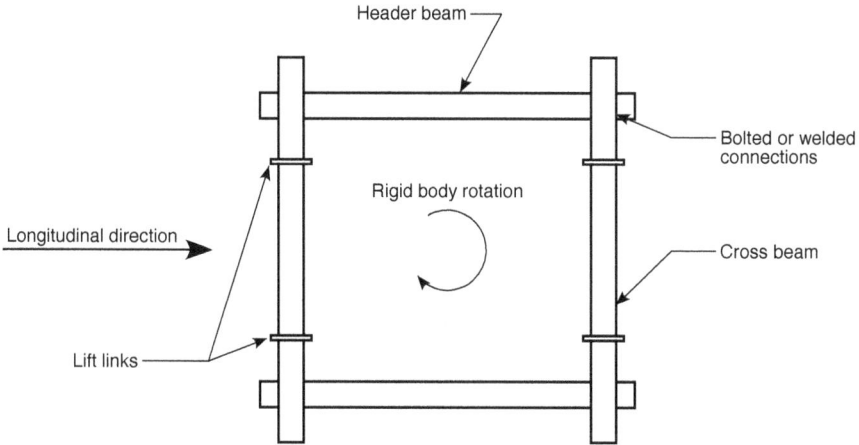

Figure 5.15 Four-Beam Arrangement Effect on Stability

beam spanning between gantry legs in the longitudinal direction and the cross beams spanning between header beams in the lateral direction. The header beam rocker is arranged such that it articulates in the longitudinal direction. Since the header beam is clamped to the header plate (as it should be), the header beam acts as an extension to the height of the gantry leg, thus increasing the height of the gantry leg with respect to the calculation of stability in the lateral direction. If the connections between the header beams and cross beams are not exceptionally rigid, then the gantry leg height H should be taken to the top of the header beam, as indicated in Fig. 5.16, for the calculation of lateral stability. If this connection is a structurally rigid connection, then the stability problem becomes one of a rigid frame analysis, rather than that of a mechanism. This is a significantly more complex problem and requires an analysis that is more sophisticated than that presented in this chapter.

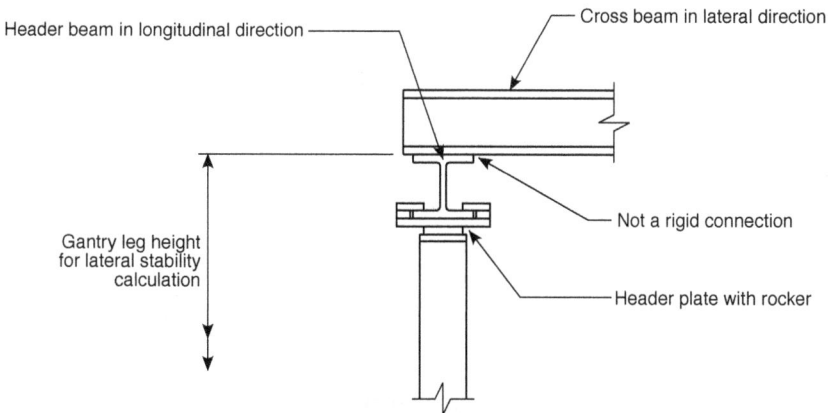

Figure 5.16 Effect of Header Plate Articulation on Stability

5.3   PRACTICAL CONSIDERATIONS

The stability analysis techniques developed in this chapter consider only the gantry system, itself, and make certain tacit assumptions. Because of this, a few final comments are in order.

The tipping fulcrum of the gantry leg for the calculation of lateral stability is taken as the outer edge of the wheel tread (Fig. 5.1). The tipping fulcrum of the gantry leg for the calculation of longitudinal stability is taken as the end axle line (Figs. 5.2 and 5.3), the end of the roller chain, or the wheel assembly pivot, as appropriate for the particular leg design. Each of these assumed fulcrum locations is based on the assumption that the gantry leg base weldment, wheel assemblies, and wheels or rollers possess the structural capacity to support the loads that will occur as an overturing force acts on the leg.

Experience in the investigation of gantry accidents has shown that this is the case. Although these components are not necessarily designed for the loads that occur during toppling, this behavior is not unexpected. Gantry toppling typically occurs only when the legs are extended to near their full height. Most gantry systems have reduced rated loads in the uppermost stage, so the base components have excess capacity relative to these lower rated loads. This excess capacity combined with the normal structural and mechanical design factors appears to provide the extra strength needed to avoid failure of these parts during a toppling incident.

Also implicit in this stability analysis is the assumption that the supporting surface can adequately resist the increased wheel loads that occur during toppling. Of particular importance when examining lateral stability is the torsional strength and stiffness of the track beams. Excessive twisting of a track beam will have the effect of moving the lateral tipping fulcrum from the edge of the wheels to the center the of the beam. The most popular types of gantry track use box shaped track beams. These beams are either fabricated box beams or wide flange beams with web plates added to form a box. Such box shaped track beams are generally strong and stiff in torsion, thus minimizing this potential support problem. Single-web track beams are typically used only with gantry models with single wheels at each corner. In this case, the offset of the loading to the track during toppling is small, as is the difference in righting arm length to the center of the beam, rather than to the edge of the wheels. A detailed discussion of gantry track design considerations and methods is found in Chapter 6.

Choices made in selecting the rigging can also affect the stability of the gantry system. Unnecessarily long slings between the lifted load and the lift links will necessitate a greater extension of the gantry legs when making the lift. All else being equal, a gantry system will be in a position of lower stability when extended to a greater height. Thus, an advantage with respect to the stability of the gantry system is realized by keeping the slings as short as is practical for the lift being made. Chapter 7 offers additional discussion of practical applications for gantry lift planning and execution.

System setup also affects the system stability, as shown by the inclusion of the track beam lateral and longitudinal slopes (quantified by the elevation differences E and S, respectively) in the calculations. Minimization of these slopes through the exercise of appropriate field controls serves to improve system stability.

5.4 REFERENCES

American Society of Mechanical Engineers (ASME) (2009), ASME B30.1-2009 *Jacks, Industrial Rollers, Air Casters, and Hydraulic Gantries*, New York, NY.

Society of Automotive Engineers, Inc. (SAE) (2007), J1078 *A Recommended Method of Analytically Determining the Competence of Hydraulic Telescopic Cantilevered Crane Booms*, Warrendale, PA.

Specialized Carriers & Rigging Association (SC&RA) (1996), *Recommended Practices for Hydraulic Jacking Systems*, Centreville, VA.

Specialized Carriers & Rigging Association (SC&RA) (2004), *Recommended Practices for Telescopic Hydraulic Gantry Systems*, Centreville, VA.

6 Engineering Methods and Practices

The engineering required as a part of the planning of a lift to be performed with a hydraulic gantry system generally does not require the use of exceptionally complex techniques. There are, however, certain calculations that must be made by the lift planner as a part of the equipment selection process. At present, the industry standards and references that address the use of hydraulic gantries provide only the most basic guidance for lift planning and engineering. Therefore, engineering decisions must often be made on the basis of the judgment and experience of the lift planner.

The typical planning of a lift using a hydraulic gantry system requires the contractor to determine the weight of the lifted load, estimate or calculate how that weight is shared among the lifting attachments (padeyes, trunnions, etc.), and to select the appropriate rigging (slings, shackles, etc.) and lift links to rig the load to the header or cross beams. The contractor must then select or design one or more header beams and cross beams, if required, from which the lifted load will be suspended. Next, calculation of the load to each gantry leg and comparison of that load to the rated load of the leg is required. Last, a track system or other supports on which the gantry legs will be carried must be selected or designed. This work may include an investigation of the strength of the soil or structures under the track system.

This chapter examines the primary engineering areas that must be addressed in the planning of a lift using a hydraulic gantry system. The intention here is to examine in depth each aspect of gantry lift engineering with which the contractor is most commonly faced, consider in a general sense the effects that each such aspect will have on the overall safety and performance of the lift, and then to offer recommendations or guidelines that the rigging contractor can apply in the planning of a gantry lift. Assumptions are discussed where appropriate in each section. Lift planning issues of a less technically intensive nature, such as determination of lift height and checking of clearances around the gantry system and the lifted load, are addressed in Chapter 7.

One more comment must be offered before delving into this subject. The discussions presented in this chapter, particularly with respect to header beam and track design, presume that the reader has a basic knowledge of structural engineering. It is not the intent of this book to teach the fundamentals of beam design. Rather, the topics covered here are those that are unique to lift engineering with hydraulic gantries and subjects that call for special explanation.

6.1   RIGGING AND LIFT LINK SELECTION

The selection of the rigging for a lift made with a gantry system generally follows the same principles that are applicable to lifts made with mobile or overhead cranes. That is, the load at each lifting attachment is determined based on the actual weight (impact factors typically are not used here) and rigging gear is selected based on the manufacturers' rated loads. There is one important difference in this process that is introduced by the way a gantry system functions. That difference regards how to deal with the potential load imbalances caused by cross cornering.

In the early days of hydraulic gantry system use, primarily the 1980s and 1990s, common practice called for sizing the rigging for a load lifted from four points based on the assumption that the entire weight would be supported at either pair of diagonally opposite corners. That is, a full 100% cross cornering was assumed. That level of conservatism may not be warranted today for two reasons. First, newer gantry systems are often equipped with control systems that continuously monitor leg extension and load and can automatically adjust the system operation to minimize cross cornering. Second, the understanding of cross cornering is more prevalent today, so contractors using older gantry systems are more likely to monitor and adjust the leg extensions to minimize cross cornering manually.

Let us examine this issue in greater detail. Consider the lift arrangement shown in Fig. 6.1. Illustrated is the very common lift of a torsionally rigid load from four lifting attachments. This figure is similar to Fig. 4.2 that was used to illustrate three examples of cross cornering. As was seen in Example 4-2, a relatively small deviation in the extensions of the gantry legs for some lift arrangements can result in a complete shifting of the weight of the lifted load to just two diagonally opposite slings. The magnitude of this load shifting is closely related to the spread of the lift links relative to the header beam span. This effect is illustrated in Fig. 6.2 for three different ratios of lift link spread to header beam span for one particular load / rigging / gantry leg arrangement. So the question before us is: Should the rigging

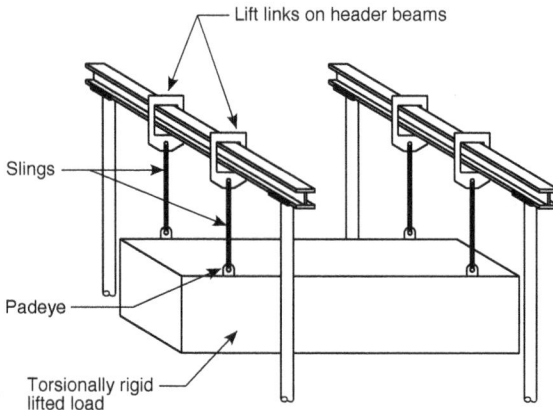

Figure 6.1 Rigging to Four Lifting Attachments

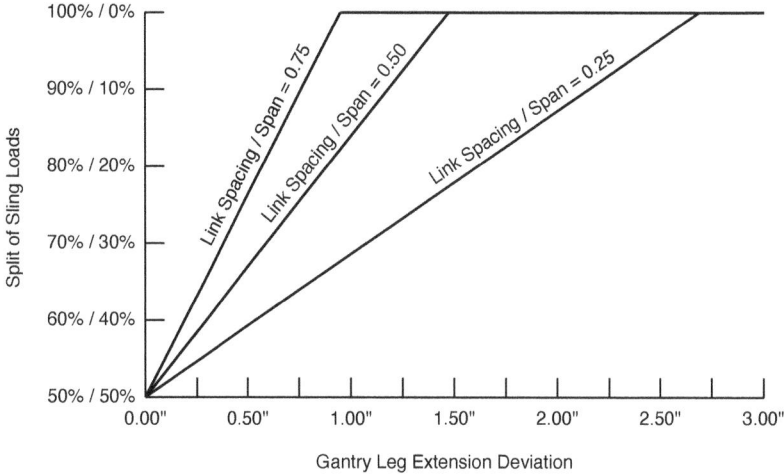

Figure 6.2 Effect of Lift Link Spread on Sling Tensions

and the lift links be selected based on the calculated load distribution assuming uniform leg extension, the worst-case loads due to full cross cornering (as was recommended in the past), or some load distribution between these two extremes?

The selection of slings, shackles, and other conventional rigging gear can normally be done based on a load distribution that assumes uniform leg extension. As discussed in Chapter 4, the effects of cross cornering can be minimized through proper monitoring and operation of the gantry system and, as noted above, gantry system manufacturers are increasingly providing control systems that continually monitor the extension of and load supported by each gantry leg and can automatically adjust the system's movement to keep the legs closely synchronized. Further, the rated loads of slings, shackles, and the like are based on safety factors of 4 or greater with respect to breaking strength, thus providing a comfortable margin against failure even when the calculated applied load is a very high percentage of the rated load and some small deviation in actual applied load occurs as a product of system operation.

Selection of the lift links is another story, however. There are presently no standards, regulations, or industry guides that prescribe strength or performance requirements for the rating of gantry system lift links. The author has seen lift link designs based on the provisions of the last allowable stress design edition of the AISC *Specification for Structural Steel Buildings* (AISC 1989). This specification provides nominal design factors of 1.67 with respect to yielding and 2.00 with respect to fracture. We can see that a lift link so designed would come very close to failure if the actual applied load was twice the rated load due to cross cornering or other operational effects. The author's suggested design method and design factor for lift links are discussed in Section 6.6.

If the rated loads of the lift links being used are based on a methodology that provides a suitable margin to protect against failure due to operationally imposed overloading, then selection of lift links based on the calculated applied load and the lift link's rated load may be acceptable. Otherwise, the lift planner should consider selecting oversized lift links or using a pair of lift links at each lift point to provide an appropriate margin.

We can see from Fig. 6.2 that the potential of overloading the rigging from cross cornering of the load is a direct function of the equipment arrangement. A setup that places the lift links close together is not very susceptible to overloading due to cross cornering. Alternately, a setup with a relatively wide lift link spread is much more susceptible to this effect. The lift planner should take this behavior into consideration when selecting rigging and lift links for a particular project.

One should also consider the gantry system being used when assessing the potential rigging load increase due to cross cornering. Is the gantry equipped with a modern control system that will monitor the leg extensions and the supported loads and automatically adjust the leg movements to minimize cross cornering? If so, the magnitude of load shifting due to cross cornering will be minor and the rigging and lift links may reasonably be selected using the calculated lifting attachment loads. If not, will other means be used to monitor the gantry system to minimize cross cornering? Even if all the gantry operator has to work with is the control unit pressure gauges and tape measures on each leg, he should be able to keep the legs reasonably synchronized, thus limiting cross cornering.

Consider, for example, the combined information shown in Figs. 6.2 and 6.3 for the wide lift link spacing of 75% of the header beam span. We see in Fig. 6.2 that one leg out of sync by 0.75 inch (19 mm) results in almost a 90% / 10% split of the load between diagonally opposite pairs of slings. We see in Fig. 6.3 that this 0.75 inch (19 mm) deviation results in an 80% / 20% split of the load between diagonally opposite pairs of gantry legs. This large deviation of the gantry leg load from the expected 50% / 50% will clearly be observed and corrected by the

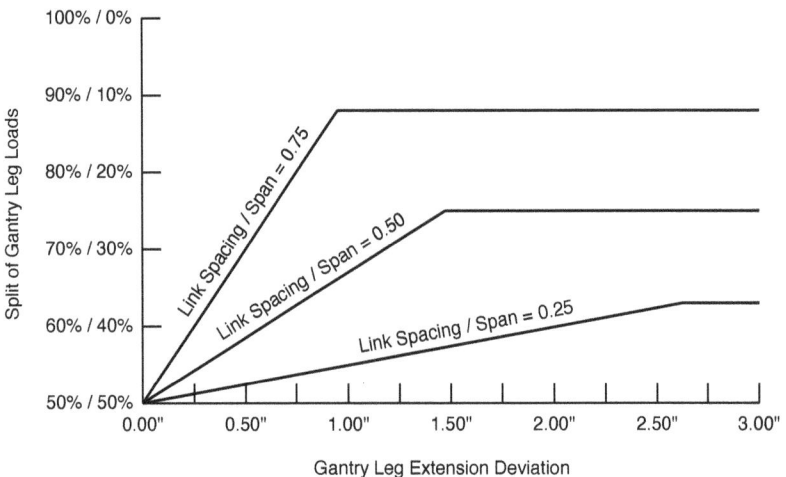

Figure 6.3 Effect of Lift Link Spread on Gantry Leg Loads

operator, most likely well before reaching this magnitude. Thus, we can see the significance of taking into account the type of gantry control system and operating practices when making rigging and lift link selections.

The longer the rigging (all else being the same), the lower its axial stiffness, thus reducing the sensitivity of the arrangement to cross cornering. However, the length of the rigging also affects the leg extension height needed to make the lift and can affect potential side loading of the gantry system if the suspended load must be drifted. The practical considerations that the lift planner must address with respect to rigging length are discussed in Section 7.1.2.

6.2   DESIGN OF HEADER AND CROSS BEAMS

Header and cross beams are designed for some combination of vertical and horizontal loads. These loads include the dead weight of the beam, the actual applied vertical load or loads, some specified vertical impact load (usually a percentage of the actual vertical load), and some specified horizontal dynamic load (also usually a percentage of the actual vertical load). The required strength of the beam is generally based on its ability to support this combination of specified loads without exceeding appropriate allowable stresses. The required stiffness of the beam is generally based on its ability to deflect no more than some specified amount, usually considering only the actual applied vertical loads. The procedure for designing a header or cross beam follows the steps listed below.

- Determine the magnitudes and positions of the loads that will act on the beam;
- Select vertical and horizontal impact factors to account for dynamic loading effects and, if appropriate, gravity load uncertainty;
- Compute the shear, bending, and local compressive stresses in the beam that result from the combined loads that act during the lift and compare those stresses to the allowable stresses for the beam;
- Calculate the support point reactions of each beam in order to carry the load from that beam through to the next element in the gantry system; and,
- Compute the vertical deflection of the beam under the effects of the actual loads and compare that value to the allowable deflection.

Each of these beam design steps is investigated in appropriate detail in the following sections.

6.2.1   Design Loads

The gravity loads that act on a header or cross beam are based on the lift link reactions to the beam, as developed in Section 6.1. These loads include the payload weight as distributed to each lift attachment and the dead weights of the rigging components and lift links.

For the most common gantry arrangement that uses only header beams, the vertical gravity loads to each header beam are the lift link reactions. For more complex beam arrangements, the gravity loads to the uppermost cross beams are the lift link reactions and the gravity loads to the lower beams (typically the header beams) are the cross beam reactions, which must include the cross beam dead weight. The location at which a lift link or a cross beam bears on the beam under consideration is called the load point. A basic header (or cross) beam with two load points is illustrated in Fig. 6.4.

The next step in the design procedure is the selection of dynamic load factors. (In this discussion, we will generally refer to all of these multipliers as "impact factors." As discussed in Chapter 4, these factors are used to account for other than true dynamic loads. The use of this term here is simply to be consistent with common industry usage.) The responsibility of determining the values of vertical and horizontal impact factors to be used in design generally rests with the engineer performing the work. The SC&RA *Recommended Practices for Telescopic Hydraulic Gantry Systems* (SC&RA 2004) can be used for guidance in selecting suitable impact factors for beam design.

Consider first the vertical impact factor. SC&RA (2004) recommends a vertical impact factor of 5% for the design of header and cross beams. This factor can be considered to address not only the (very) minor vertical dynamic loading that may occur in routine lifting, but also to provide a margin of safety with respect to inaccuracies in the weight of the lifted load. A larger vertical impact factor may be considered if the operation involves unusual movement that could result in significant dynamic loading. One potentially significant source of vertical impact is the jolt that can occur when using gantries to downend a load.

As discussed in Chapter 4, a downending operation should never be planned with the intention of letting the load roll over center and then "catching" the load with the gantries. However, an error in the determination of the center of gravity of the load can inadvertently result in unexpected movement of the load or a

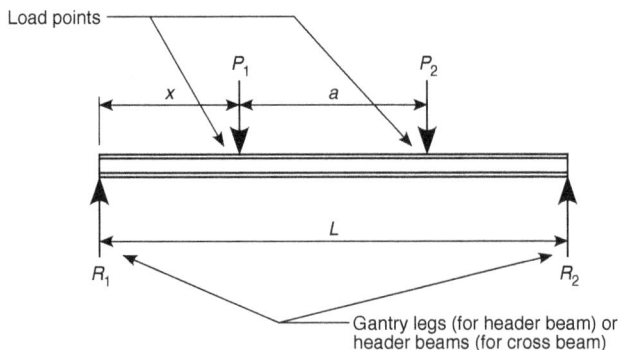

Figure 6.4 Vertical Loads on a Header or Cross Beam

distribution of weight between gantries that varies from the computed distribution. Therefore, a greater vertical impact factor may be appropriate for header beam design when using the gantry system to downend a load. In the past, the author has suggested a vertical impact factor of 15% to 20% for this situation. The actual factor to be used for any particular lift must be determined by the engineer based on the specific circumstances.

Header and cross beam design must also consider the horizontal forces that may occur during the lift. SC&RA (2004) suggests that header and cross beams be designed using a horizontal impact factor of 5% for a straight lift or 10% for a lift with travel. The discussions of horizontal loads in Chapter 4 showed a maximum individual load of about 1.5% of the vertical load. Reasonable combinations of horizontal loads, although not developed numerically in Chapter 4, may be somewhat greater, but most likely not much greater. Thus, the comparatively high values suggested in SC&RA (2004) at first appear to be unreasonable. However, when taken in combination with common beam design practices used in the industry, these values are usually quite reasonable. Let us examine why this is so.

It is common practice in the heavy rigging industry to design header and cross beams based on the assumption that the vertical and horizontal loads act through the centroidal axes of the beam. This is usually appropriate for the vertical load at each load point, but not normally so for the horizontal load. In reality, the horizontal load F is applied at some distance below the bottom of the beam when the rigging is attached to the beam by means of a lift link, as illustrated in Fig. 6.5. Thus, the locations of the applied loads are such that the beam is subjected to a major axis moment (M_x), a minor axis moment (M_y), and a torsion (T), not just the two moments.

To better understand the effect that this torsion has on the strength of the beam, consider a simplified treatment of a wide flange beam subjected to torsion. A torsion T acting on the beam can be resolved into two equal but opposite forces acting at each flange. The beam height from center of flange to center of flange is taken as h' and the center of the lift link pin below the center of the bottom flange

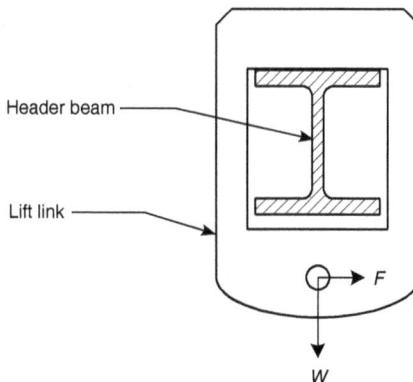

Figure 6.5 Loads Acting at the Lift Link Pin Hole

is taken as h_l. The flange forces are then computed as follows. Refer to Fig. 6.6 for an illustration of this process.

$$T = F\left(h_l + \frac{h'}{2}\right) \tag{6.1}$$

$$P_u = \frac{T}{h'} - \frac{F}{2} = F\left(\frac{h_l}{h'} + \frac{h'}{2h'} - \frac{1}{2}\right)$$

$$P_u = F\frac{h_l}{h'} \tag{6.2}$$

$$P_l = \frac{T}{h'} + \frac{F}{2} = F\left(\frac{h_l}{h'} + \frac{h'}{2h'} + \frac{1}{2}\right)$$

$$P_l = F\left(\frac{h_l}{h'} + 1\right) \tag{6.3}$$

Each force P_u or P_l is treated as a horizontal load that creates a moment in the beam, but that moment is resisted by only half of the beam's minor axis section modulus. As illustrated in Fig. 6.7, the force P_l creates a minor axis moment and the bending stress from this moment is computed using only one-half of S_y. Using this simplified method of analyzing a wide flange beam subjected to torsion, we must now look at the magnitude of that torsion in the design of gantry system header and cross beams.

The typical lift link proportions provide a few inches of clearance between the bottom flange of the header or cross beam and the bottom of the beam opening in the lift link. The depth of the bottom segment of the link and the position of the pin hole within that segment vary, depending on the rated load of the lift link. Considering typical beam sizes and lift link designs, we can reasonably assume that the distance from the bottom flange of the header beam to the pin hole will not exceed two-thirds of the center-to-center distance between the beam flanges. That is, we may take as an upper bound value of the height from the center of the bottom flange of the beam to the center of the lift link pin hole h_l as $0.67h'$. This allows us to rewrite Eqs. 6.1 and 6.3 as follows for this presumed upper bound case.

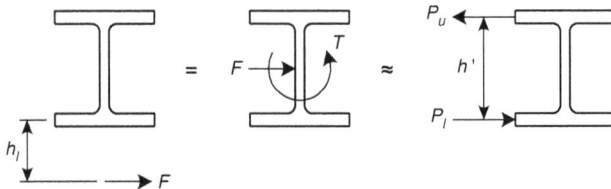

Figure 6.6 Eccentric Horizontal Load Acting on a Wide Flange Beam

$$T = F\left(0.67h' + \frac{h'}{2}\right) = F\left(1.17h'\right) \tag{6.4}$$

$$P_l = F\left(\frac{0.67h'}{h'} + 1\right) = 1.67F \tag{6.5}$$

Let us now consider a beam carrying a vertical load of 100 kips. The horizontal load F is taken as 1.5% of this vertical load, or 1.5 kips. We can used the relationships illustrated here to develop the flange forces that result from this horizontal load.

The left diagram in Fig. 6.6 shows the actual location of the horizontal load F. This can be resolved into a horizontal load acting at the centroid of the beam plus a torsion T (center diagram), where T is calculated using Eq. 6.4. Last, we can approximate this loading by resolving the torsion into a pair of flange forces (right diagram), to which is added (for the lower flange) one-half of force F (Eq. 6.5).

If the actual horizontal load F equals 1.5 kips, then the upper bound of the resulting flange force P_l is found to be 2.5 kips from Eq. 6.5. As noted above, the minor axis bending stress due to the moment created by this flange force is calculated using one-half of the minor axis section modulus. Therefore, if we use double this flange force, or 5.0 kips (which is 5% of the vertical load), and the full minor axis section modulus to calculate the minor axis bending stress, we effectively approximate the minor axis bending stress that will result from a horizontal load of 1.5 kips acting at the upper bound eccentricity of F developed above.

Thus, we see from this analysis that use of a horizontal impact factor of 5% in conjunction with the common practice of treating the resulting horizontal force as acting through the centroid of the beam produces approximately the same load effect as a horizontal load equal to about 1.5% of the vertical load acting at its true eccentric location. Obviously, the upper SC&RA (2004) value of 10% for lifts with travel equates to a horizontal load of 3.0% of the vertical load again acting

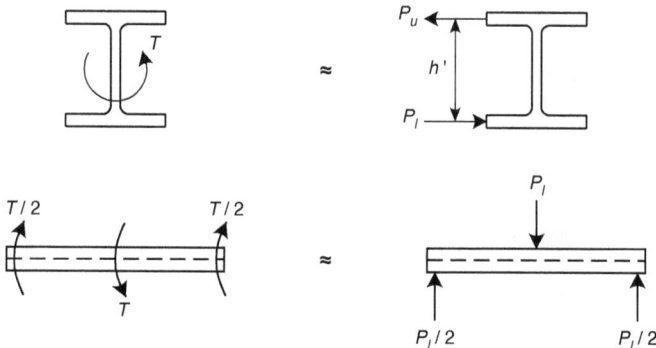

Figure 6.7 Torsion Analogy for a Wide Flange Beam

at its true location. Given the magnitudes of the horizontal loads developed in Chapter 4, the SC&RA (2004) value of 10% is somewhat conservative.

The beam and lift link proportions illustrated in Fig. 6.5 are based on typical jumbo wide flange beams. We can immediately see that the torsion and flange force derivation discussed on the preceding pages will vary, possibly significantly so, for beams and lift links with different proportions. Consider, for example, a W36 header beam with a depth of about 36 inches (915 mm). The height from the center of the bottom flange to the center of the lift link pin hole may be on the order of 10 to 12 inches (250 to 300 mm). Letting $h' = 34.5"$ and $h_l = 12"$, we can solve Eq. 6.3 to obtain $P_l = 1.35 \, F$. Thus, the design horizontal load of 10% of the vertical load applied through the centroid of the beam is equivalent to a horizontal load of 10% / 1.35 / 2 = 3.7% acting at the lift link pin hole. Thus, we see that this simplified loading model may be very conservative relative to the expected magnitudes of the horizontal loads for some beam / lift link proportions.

Another deviation occurs when the header or cross beam is a box section, rather than a wide flange beam. The torsional strength and stiffness of a closed shape is typically significantly greater than that of an I-shape member of similar bending strength, so the torsion analogy illustrated in Figs. 6.6 and 6.7 and upon which Eqs. 6.1 through 6.5 are based will yield very conservative results.

And last, the torsion in a header beam developed by a horizontal force will be that much less when the horizontal force is a reaction from a cross beam mounted on the top flange of the header beam. In this circumstance, the moment arm through which the horizontal force acts is only one-half of the beam depth.

In spite of the very conservative results that may occur with certain beam types and proportions, the simple provisions given in SC&RA (2004) are practical in that they produce reasonable results for the most commonly used jumbo wide flange beams and lift link proportions. This approach also allows for the development of header beam load charts, which can be very useful for basic lift planning, on which the rated loads are not dependent on the proportions of the lift links used or load application from a cross beam, rather than a lift link.

In summary, the values of 5% vertical impact plus 5% horizontal impact for straight vertical lifts and values of 5% vertical impact plus 10% horizontal impact if the load is to be traveled in a direction perpendicular to the span of the beam as suggested in SC&RA (2004) are considered appropriate for most gantry system lifts. Experience has shown that these values produce safe, but economical, beam designs. However, the responsible engineer must always consider the actual conditions of the lift being planned and revise these impact values if appropriate.

6.2.2   Shear, Moments, and Reactions

With the loads determined and suitable impact factors selected, the maximum bending moments (major axis and minor axis), shear, and support reactions can be calculated for each cross beam and each header beam. These calculated (or demand) values are then compared to suitable allowable (or capacity) values to determine the adequacy of the beams.

Current practice in gantry lift engineering employs the methods and allowable stresses of the AISC *Specification for Structural Steel Buildings* for header beam and cross beam design. SC&RA (2004) cites the last edition of the allowable stress design version of this specification (AISC 1989), but use of the allowable strength provisions of the current version (AISC 2010b) is also reasonable, if not preferable. The AISC allowable stresses are based on a nominal design factor of 1.67 with respect to yield stress or buckling strength. The various provisions of the *Specification* define formulas for checking bending, shear, connections, buckling, and web crippling; i.e., every aspect of beam design. This gives us a comprehensive and convenient means of obtaining a safe beam design.

The design of a header or cross beam on which the load points are fixed is very simple. The loads and load layout as illustrated in Fig. 6.4 are used to calculate the shear and moments using simple statics. When using AISC (1989), the shear and moments are used to calculate shear and bending stresses. These actual stresses are then compared to the allowable stresses prescribed by the specification, using the appropriate interaction formula for the combination of major and minor axis bending. When using AISC (2010b), the specification defines allowable shear and moments, rather than allowable stresses, so the intermediate calculation of actual stresses is not necessary.

If the lifted load is to be side-shifted along the beam, the limits of that travel must be defined. Under most circumstances, the beam calculations will have to be performed for three positions of the load: left, middle, and right (Fig. 6.8). The leftmost position (Fig. 6.8a) gives the maximum reaction and shear at the left end of the beam. The rightmost position (Fig. 6.8c) does the same to the right end of the beam. The maximum bending moments occur with the moving loads positioned near the center of the span (Fig. 6.8b).

When the two loads on the beam are equal, the maximum moment occurs under load P_1 when the locating dimension x_2 is as given by Eq. 6.6. The reader may note that structural engineering references usually state that this is true only when the load spacing a is less than 0.586 L. When a is greater than 0.586 L, then the maximum moment occurs with one load centered on the span and the other load off the span. Since such a position is not normally possible with a gantry system header and cross beam arrangement, this second possible case does not occur and need not be considered.

$$x_2 = \frac{1}{2}\left(L - \frac{a}{2}\right) \tag{6.6}$$

When load P_1 is greater than P_2, then the locating dimension x_2 is calculated using Eq. 6.7.

$$x_2 = \frac{1}{2}\left(L - \frac{P_2 a}{P_1 + P_2}\right) \tag{6.7}$$

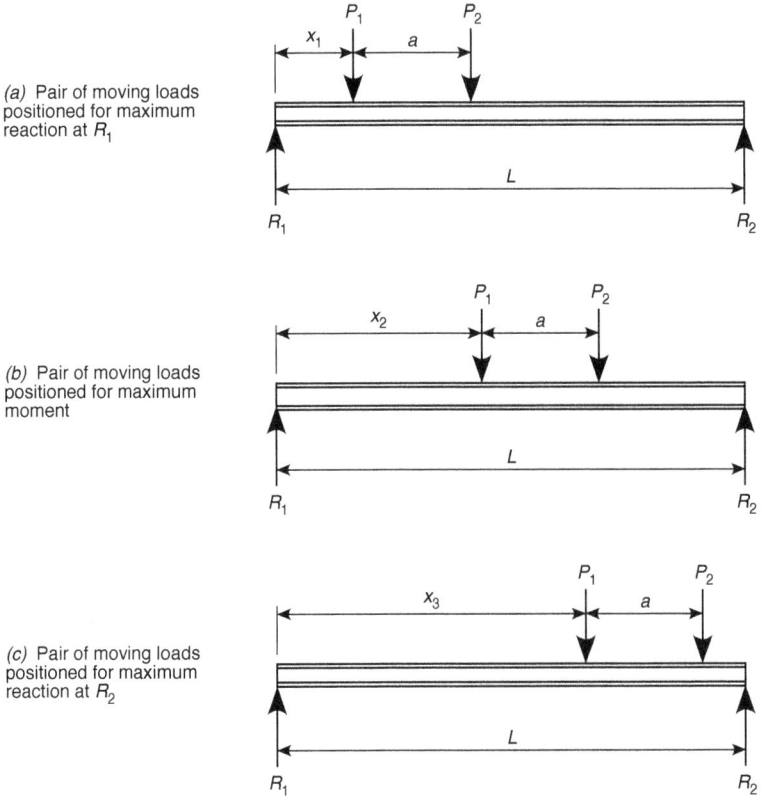

Figure 6.8 Moving Loads on a Header or Cross Beam

The design of a beam loaded by three or more moving loads requires a more detailed solution. In general, the maximum moment occurs under one of the loads when that load is as far from one support as the center of gravity of all of the loads is from the opposite support. Of course, the position of the group of moving loads at which the maximum moment occurs need not move outside of the envelope defined by the limits of movement that will occur for the lift under consideration (dimensions x_1 and x_3 in Fig. 6.8).

As with the rigging selection, the effect of cross cornering must be considered in the design of the header and cross beams. The effect of cross cornering on the beam's shear is the same as the effect on the gantry leg load (Fig. 6.3). The effect of cross cornering on the bending moment is likewise the same. The increase in moment relative to the magnitude of cross cornering is plotted in Fig. 6.9 for three different link spacing to span ratios, again using the load and rigging properties discussed in the previous examples. We can see, for example, that at a link spacing to span ratio of 0.50, full cross cornering results in the loads to the more heavily loaded gantry legs increasing by about 50% (from Fig. 6.3), and the beam moment increases by the same 50% (Fig. 6.9).

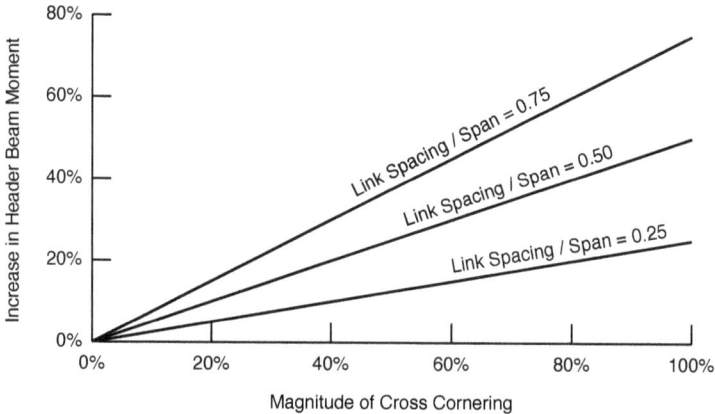

Figure 6.9 Cross Cornering Effect on Bending Moment

Last, we must compute the reaction at each beam support. Header beam reactions are calculated using only the gravity loads. These are the lift link loads applied to the beam and the dead weight of the beam, itself. Two sets of reactions must be calculated for cross beams. The first set are the beam reactions based on the applied loads including all impact factors. These values are used as the loads applied to the header beams. The second set are the beam reactions considering only gravity loads. These values will be carried through for the calculation of the header beam reactions due to gravity loads only.

6.2.3   The Effects of Concentrated Loads

The concentrated loads applied to the top flange of a header or cross beam through the lift links or side shifting devices create local compressive stresses in the beam web. These stresses can cause local yielding of the web, crippling of the web, or lateral buckling of the tension (lower) flange and must be checked as a part of the design of the beam.

The AISC specifications (AISC 1989, AISC 2010b) provide methods of evaluating the strength of a web under the effect of a single concentrated load. However, many of the side shift devices in use today have two or three axles relatively close to one another. This loading requires modifying the standard web strength evaluation methods. Suitable formulas are presented here.

Web Local Yielding. The web local yielding strength is based on the assumption that the applied load spreads out over a length of the web equal to the width of the applied load plus five times the height from the surface of the flange to the bottom of the web-to-flange fillet (for rolled shapes) or web-to-flange weld (for built-up shapes). The development of this design provision can be found in Graham, et al

(1959). The web yielding strength R_{wy} can be expressed as Eq. 6.8.

$$R_{wy} = F_{yw}t_w\left(5k + l_b\right)\big/1.50 \qquad (6.8)$$

where

F_{yw}	=	yield stress of the web;
t_w	=	web thickness;
k	=	distance from the outer face of the flange to the web toe of the web-to-flange fillet or the web-to-flange weld; and,
l_b	=	length of bearing of the concentrated load ($l_b = 0$ for a wheel or roller).

When working with older references, note that the letter N was used to indicate the length of bearing in earlier editions of the AISC specification and in past publications on this subject.

The divisor of 1.50 in Eq. 6.8 is the design factor prescribed by AISC (2010b). The web yielding provision in AISC (1989) uses a multiplier of 0.66, rather than a divisor of 1.50, which produces a nominally identical result. As discussed above with respect to other aspects of beam design, the use of the AISC design factors is generally appropriate for gantry system beam design.

Now consider the two arrangements shown in Fig. 6.10. Eq. 6.8 is based on the loading shown in Fig. 6.10a. The loading shown in Fig. 6.10b is the common (in gantry applications) situation of two equal concentrated loads in close proximity. If the two loads are too close to one another, the stress footprints upon which Eq. 6.8 is based will overlap. More specifically, if the distance s between the two loads is less than $5k$, then web local yielding should be evaluated using Eq. 6.9. In this usage, the web yielding strength R_{wy} is compared to the total applied load $2P$.

$$R_{wy} = F_{yw}t_w\left(5k + 2l_b + s\right)\big/1.50 \qquad (6.9)$$

If the distance s between the two loads is equal to or greater than $5k$, then the loads act independently and web local yielding may be evaluated using the normal AISC provision (Eq. 6.8).

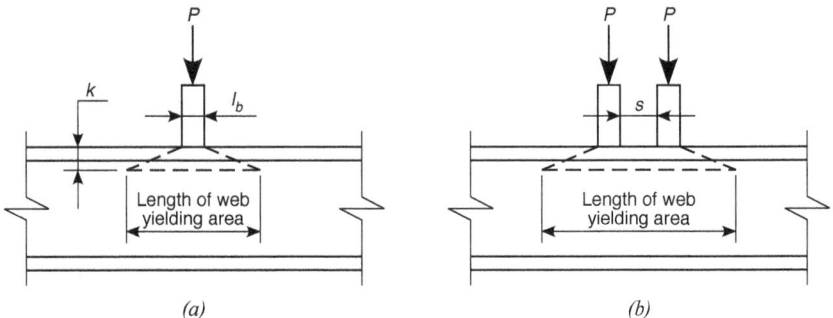

Figure 6.10 Load Positions for Check of Web Local Yielding

Web Local Crippling. The web local crippling strength limit state is a condition in which the beam web fails as a result of the formation of plastic hinges in the flange and the web (Fig. 6.11). The provisions in AISC (1989) are based on Roberts (1981). The provisions in AISC (2010b) also consider Elgaaly and Salker (1991) for the condition in which the distance from the end of the beam to the concentrated load is less than one-half of the beam depth.

The model upon which the AISC provisions are based is shown in Figs. 6.11*a* and 6.11*b*. At the limit state load, plastic hinges form in the flange at either edge of the body through which the concentrated load is applied and at a distance β to either side. Further, hinges form in the web along the web-flange juncture and along two lines as illustrated.

When two or more loads are supported on the beam at a great enough distance from one another, then the AISC provisions can be applied without modification. When two loads are close enough that the regions defined in the failure model interact, then a modification is needed. The limiting clear distance between the loads may be taken as L_{cr} as defined in Eq. 6.10 [1].

$$L_{cr} = \frac{d}{3}\left(\frac{t_f}{t_w}\right)^{1.5} \tag{6.10}$$

where
d = beam depth;
t_f = flange thickness; and,
all other terms are as previously defined.

When s is equal to or greater than L_{cr}, then the concentrated loads can be treated

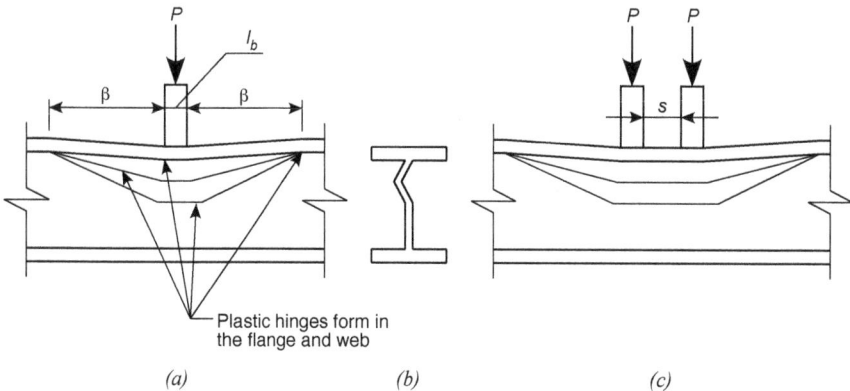

Plastic hinges form in the flange and web

(a)   (b)   (c)

Figure 6.11 Load Positions for Check of Web Local Crippling

---

[1]The author wishes to thank Bo Dowswell, Ph.D., P.E. of SDS Resources, LLC, Birmingham, Alabama, for the development of Eq. 6.10.

independently and the normal AISC provisions are applied. When s is less than L_{cr}, then the two loads act together as illustrated in Fig. 6.11c. In this case, one total limit state load is calculated using the AISC equations (either AISC 1989 or AISC 2010b as appropriate) using a width of bearing equal to $2l_b + s$.

Web Sidesway Buckling. The last beam strength limit state related to the effects of concentrated loads that bears special investigation is that of web sidesway buckling (Fig 6.12). This failure is a buckling of the web of an I-shape beam accompanied by a sideways shifting and twisting of the bottom (tension) flange of the beam due to the application of a concentrated load on the top (compression) flange. The AISC provisions, which are based on Summers and Yura (1982), again consider only a single concentrated load acting on the beam.

A study of web sidesway buckling due to both single concentrated loads and pairs of loads has been performed by the author using BASP (Buckling Analysis of Stiffened Plates), a finite element analysis program developed at the University of Texas at Austin for the solution of elastic buckling problems. (For reference, BASP was also used in the original Summers and Yura 1982 work.) The results of this study show that two concentrated loads can be treated as independent when the spacing s from center to center of the loads is at least 2.50 times the clear web height h. If the spacing s between loads is less than 2.50 times the web height h, then the allowable web sidesway buckling load R given by the appropriate formula in either AISC (1989) or AISC (2010b) must be reduced by the correction factor c given in Eq. 6.11.

$$c = 0.5 + \frac{s}{5h} \leq 1.0 \tag{6.11}$$

One will notice significant differences in the web sidesway buckling equations in AISC (2010b) verses AISC (1989). The AISC (1989) equations are simplified for beams that have proportions more in line of those shapes most commonly used in building construction. When used to check beams of the stout proportions often used as gantry header and cross beams, the AISC (1989) equations often produce

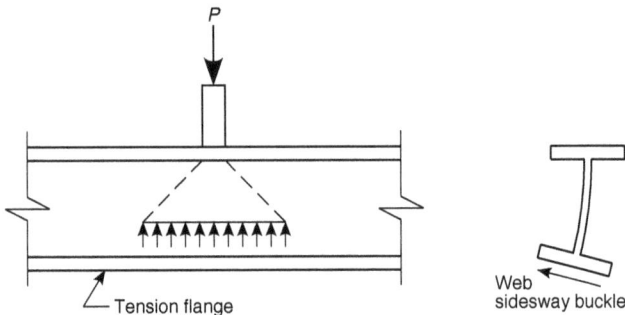

Figure 6.12 Web Sidesway Buckling in I-Shape Beams

excessively conservative results. The AISC (2010b) equations are more general and, as such, produce results that are consistent across a wide range of beam proportions. Thus, the use of AISC (2010b) is preferable. Since design standards should not be mixed, this tells us that AISC (2010b) is usually the better choice for beam design whenever the option is available.

The discussion in this section is directed to the design of header beams or cross beams following the requirements of either AISC (1989) or AISC (2010b). It is well understood that some gantry users, particularly those located outside of the United States, may apply other design standards. In such a case, it is incumbent upon the lift planner / engineer to understand the bases of the provisions of the design standard being used and to apply suitable modifications to those provisions if the application to gantry system lift engineering is not consistent with the principles that underlie that design standard.

Eqs. 6.9, 6.10, and 6.11 are based on the assumption that the two concentrated loads are approximately equal. In any case where this condition is not true, the engineer must adjust the design method appropriately.

6.2.4   Beam Design for an Underhung Trolley

It is occasionally convenient to mount a hoist on a header or cross beam using an underhung trolley. The wheel loads from such an installation produce local stresses in the bottom flange of the beam that must be calculated and combined with the bending stresses due to major and minor axis bending to evaluate the adequacy of the beam. The equations presented here are based on the provisions of CMAA (2010b) and FEM (1983). These provisions were derived from the results of an experimental investigation of the behavior of beam flanges subjected to concentrated loads that is reported in Hannover and Reichwald (1982).

Local bending stresses due to the wheel loads must be calculated at four points on the bottom flange. These are at the outer fiber in line with the face of the web (Point 0), at the outer fiber on the centerline of the contact between the wheel and the flange (Point 1), at the outer fiber at the edge of the flange (Point 2), and at the flange-to-web transition (Point 3). These points and relevant dimensions that will be used in the calculations are illustrated in Fig. 6.13 for the common case of an I-shape beam and trolley wheels that have their treads tapered to match the slope of the beam flange surface.

Eight local stresses are calculated. These are x and y direction stresses at each of the four points. In these calculations, the x direction is perpendicular to the longitudinal axis of the beam and the y direction is along its longitudinal axis. The eight local stresses are calculated using Eqs. 6.12 through 6.19 where P is the wheel load. Positive stress values are tensile and negative values are compressive.

$$\sigma_{X0} = C_{X0}\frac{P}{\left(t_a\right)^2} \qquad (6.12)$$

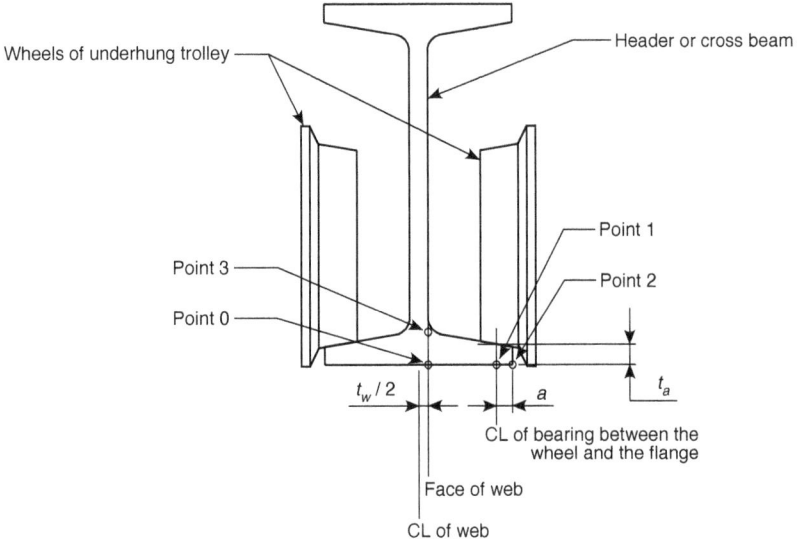

Figure 6.13 Wheel Loads and Stress Points with an Underhung Trolley

$$\sigma_{Y0} = C_{Y0} \frac{P}{(t_a)^2} \tag{6.13}$$

$$\sigma_{X1} = C_{X1} \frac{P}{(t_a)^2} \tag{6.14}$$

$$\sigma_{Y1} = C_{Y1} \frac{P}{(t_a)^2} \tag{6.15}$$

$$\sigma_{X2} = C_{X2} \frac{P}{(t_a)^2} \tag{6.16}$$

$$\sigma_{Y2} = C_{Y2} \frac{P}{(t_a)^2} \tag{6.17}$$

$$\sigma_{X3} = -\sigma_{X0} \tag{6.18}$$

$$\sigma_{Y3} = -\sigma_{Y0} \tag{6.19}$$

The flange thickness t_a used in Eqs. 6.12 through 6.17 is the thickness at Point 1 and is calculated using Eq. 6.20 for beam sections with tapered flanges, such as I shapes, where b_f is the flange width. When analyzing a beam with parallel flange surfaces, such as a W shape, t_a is equal to the specified flange thickness t_f.

$$t_a = t_f - \frac{b_f}{24} + \frac{a}{6} \tag{6.20}$$

The use of tapered wheels on a tapered flange section or straight wheels on a parallel surface flange section is, of course, preferred. However, there are instances where a trolley with straight wheels will be used on a tapered flange section or tapered wheels will be used on a parallel surface flange section. These conditions are illustrated in Fig. 6.14. These two conditions simply alter the location of Point 1 and, correspondingly, the value of dimension a.

If either of the arrangements shown in Fig. 6.14 are used, one must realize that the relatively high contact stresses between the wheels and the flange likely will cause permanent deformations in both the wheel treads and the flange. Although these deformations will not diminish the load carrying capacity, they may cause problems during future use of the equipment. Thus, these arrangements should be avoided unless absolutely necessary.

The coefficients C used in Eqs. 6.12 through 6.17 are calculated using Eqs. 6.21 through 6.26 for tapered flange sections and with Eqs. 6.27 through 6.32 for parallel flange sections. In all cases, λ is calculated using Eq. 6.33.

$$C_{X0} = -1.096 + 1.095\lambda + 0.192e^{-6.000\lambda} \tag{6.21}$$

$$C_{X1} = 3.965 - 4.835\lambda - 3.965e^{-2.675\lambda} \tag{6.22}$$

$$C_{X2} = 0.000 \tag{6.23}$$

$$C_{Y0} = -0.981 - 1.479\lambda + 1.120e^{1.322\lambda} \tag{6.24}$$

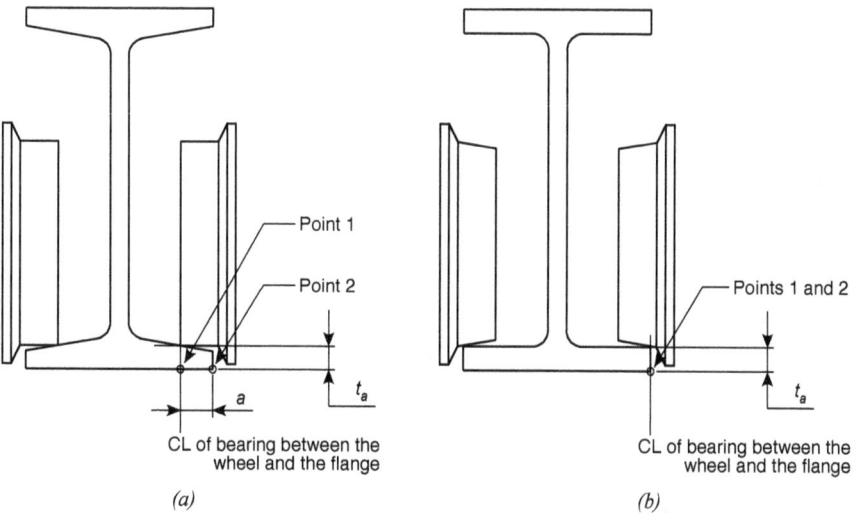

Figure 6.14 Wheel Loads and Stress Points with an Underhung Trolley

$$C_{Y1} = 1.810 - 1.150\lambda + 1.060e^{-7.700\lambda} \tag{6.25}$$

$$C_{Y2} = 1.990 - 2.810\lambda - 0.840e^{-4.690\lambda} \tag{6.26}$$

$$C_{X0} = -2.110 + 1.977\lambda + 0.0076e^{6.530\lambda} \tag{6.27}$$

$$C_{X1} = 10.108 - 7.408\lambda - 10.108e^{-1.364\lambda} \tag{6.28}$$

$$C_{X2} = 0.000 \tag{6.29}$$

$$C_{Y0} = 0.050 - 0.580\lambda + 0.148e^{3.015\lambda} \tag{6.30}$$

$$C_{Y1} = 2.230 - 1.490\lambda + 1.390e^{-18.330\lambda} \tag{6.31}$$

$$C_{Y2} = 0.730 - 1.580\lambda + 2.910e^{-6.000\lambda} \tag{6.32}$$

$$\lambda = \frac{a}{b' - t_w/2} \tag{6.33}$$

where
b' = $b_f/2$ for symmetrical single-web shapes;
  = centerline of web to edge of flange for other shapes; and,
    all other terms as previously defined.

It is obvious that the wheel loads also develop shear stress in the through thickness direction in the lower flange of the beam. Neither the CMAA (2010b) nor the FEM (1983) equations provide for the calculation of this shear stress and experience using other beam analysis methods shows that the magnitude and location of this shear stress is not significant.

These local stresses in the lower flange are combined with the major and minor axis bending stresses as follows. The factor of 0.75 by which the local stresses are reduced is as specified by both CMAA (2010b) and FEM (1983). In keeping with the coordinate system defined for the local stresses, we will let f_Y equal the total normal stress acting parallel to the longitudinal axis of the beam. Each subscript will be expanded with a 0, 1, 2, or 3 to indicate the point at which stresses are being calculated.

$$f_{Y0} = \frac{M_x c}{I_x} + \frac{M_y t_w}{2 I_y} + 0.75\sigma_{Y0} \tag{6.34}$$

$$f_{Y1} = \frac{M_x c}{I_x} + \frac{M_y (b_f/2 - a)}{I_y} + 0.75\sigma_{Y1} \tag{6.35}$$

$$f_{Y2} = \frac{M_x c}{I_x} + \frac{M_y b_f}{2 I_y} + 0.75 \sigma_{Y2} \tag{6.36}$$

$$f_{Y3} = \frac{M_x y}{I_x} + \frac{M_y t_w}{2 I_y} + 0.75 \sigma_{Y3} \tag{6.37}$$

The critical stress f_{cr} at each of the four indicated points is calculated by combining these longitudinal direction stresses with the transverse direction stresses. Based on the provisions given in CMAA (2010b) and FEM (1983), this is accomplished using the Huber-vonMises Energy of Distortion Theory formula (Eq. 6.38).

$$f_{cr} = \sqrt{\left(\sigma_X\right)^2 + \left(f_Y\right)^2 - \sigma_X f_Y} \tag{6.38}$$

where

M_x	=	major axis moment;
M_y	=	minor axis moment;
I_x	=	major axis moment of inertia;
I_y	=	minor axis moment of inertia;
c	=	distance from the major axis neutral axis to the bottom surface;
y	=	distance from the major axis neutral axis to Point 3; and,
		all other terms are as previously defined.

The critical stress f_{cr} so calculated at each of the four points is compared to the allowable stress F_{cr}. Given the allowable stress provisions of AISC (1989), AISC (2010b), and CMAA (2010b), an appropriate allowable stress is $F_{cr} = 0.66 F_y$. The combined major axis and minor axis bending strength of the beam must also be evaluated by the standard beam design methods (e.g., AISC 2010b).

The last local stress that may be evaluated is that of bearing between the wheels and the surface of the beam flange. Neither FEM (1983) nor CMAA (2010b) prescribe an allowable bearing stress for a flat surface loaded by a wheel and experience shows that this is rarely, if ever, a concern when using an underhung trolley on a gantry system header beam. However, if such a check is required, Eqs. 6.39 through 6.42, adapted from AISC (2010b), can be used. These equations are dimensionally dependent, where Eqs. 6.39 and 6.41 are for use with USCU and Eqs. 6.40 and 6.42 are for use with SI units.

$$R_n = 0.030\left(F_y - 13\right) l_b d \quad \text{(when } d \le 25 \text{ inches)} \tag{6.39}$$

$$R_n = 0.030\left(F_y - 90\right) l_b d \quad \text{(when } d \le 635 \text{ mm)} \tag{6.40}$$

$$R_n = 0.150\left(F_y - 13\right) l_b \sqrt{d} \quad \text{(when } d > 25 \text{ inches)} \tag{6.41}$$

$$R_n = 0.755\left(F_y - 90\right)l_b\sqrt{d} \quad \text{(when } d > 635 \text{ mm)} \tag{6.42}$$

where

R_n = allowable wheel load, in kips or newtons;
F_y = beam flange yield stress, in ksi or MPa;
l_b = width of wheel tread bearing on the flange, in inches or mm; and,
d = wheel diameter, in inches or mm.

Eqs. 6.39 through 6.42 apply to the use of a tapered tread wheel on a tapered flange or a flat tread wheel on a parallel surface flange beam. These equations are not applicable to either of the configurations shown in Fig. 6.14.

Although these equations as given in AISC (2010b) are described as applicable to expansion rollers and rockers, the underlying research (Wilson 1929, Wilson 1934) indicates that they are reasonably applied to slow-rolling applications as we have in this usage.

The last design issue to be addressed applies to fabricated beams. The design of the flange-to-web welding of a built-up beam, either single-web or box shape, must consider the vertical shear developed in the lower welds by the wheel loads acting on the lower flange, along with the usual horizontal shear load. Experience shows that the use of an underhung trolley on a built-up gantry system header beam is very unusual, so the calculation methods needed for this check will not be expanded upon here. CMAA (2010b) can be used to provide additional guidance for such a design.

6.2.5   Header and Cross Beam Deflections

The need to limit the deflections of header beams and cross beams is not entirely agreed upon within the industry. One gantry system manufacturer recommends that the mid-span deflection of header and cross beams be limited to 1/2 inch in 20 feet (10.4 mm in 5 meters), which is equal to Span / 480. Others experienced in the design and use of hydraulic gantry systems do not feel that specific header and cross beam deflection limits are needed for routine use. SC&RA (2004) only directs the lift planner to the gantry system manufacturer for guidance on header and cross deflection limits. One issue that is agreed upon is that when beam deflections are calculated, the calculations are performed using only the gravity loads supported by the beam. That is, the loads used in the deflection calculations need not include impact multipliers.

The significance of limiting the deflections of the header and cross beams depends somewhat on the overall lifting operation. If the lifted load is to be side-shifted along the beam, particularly when using an internally powered wheel-mounted side shifting device, then deflection limits may be appropriate. This condition is analogous to the deflection limits required in the design of runway beams for overhead bridge cranes and monorails. For example, CMAA (2010a) requires that the deflection of overhead crane runway beams not exceed Span / 600, or 0.40 inch per 20 feet of span (8.3 mm per 5 meters of span). CMAA (2010b) requires that the deflection of runway beams for under running cranes not exceed Span / 450, or 0.53 inch per 20 feet of span (11.1 mm per 5 meters of span). The reasoning here is obvious. If the beam is allowed to deflect excessively, a greater degree of mechanical power is needed to be sure the moving load does not run "downhill" uncontrolled toward mid-span and to climb "uphill" when moving away from mid-span.

The flexibility of the header and cross beams also affects the magnitude of vertical impact force that is developed as either the suspended load is side shifted or the entire gantry system is traveled. The engineering details of this behavior is discussed in Section 4.1.4. However, the real world effect of the header beam or cross beam bending stiffness, as quantified by the beam's major axis moment of inertia I_x and the material's modulus of elasticity E, is trivial, as demonstrated in Example 4-5. Thus, header and cross beam deflection limits do not contribute meaningfully to controlling elasticity-based impact loads.

Vertical deflection of the header or cross beam is also related to the rotations of the ends of the beam. Limiting the deflection of the beam is a convenient method of limiting the rotations of the ends of the beams at the supports. Consider, for example, the previously discussed four-beam arrangement. As the ends of the cross beams rotate due to the applied loads, torsion is introduced in the header beams. The stiffer the cross beams, the less will be these end rotations and, correspondingly, the less will be this torsion.

The mid-span deflection for a beam that is supporting a pair of moving loads that are equal to one another, as may occur when a header or cross beam is supporting a pair of loads that is being side shifted, is generally at its maximum value when the loads are centered on the span, rather than at the offset position that produces the maximum bending moment. When the moving loads are not equal or there are three or more loads in the group, calculation of the deflection are two or more load positions may be necessary to assure determination of the maximum value.

Once the mid-span deflection has been calculated, the question remains: What is an acceptable deflection? As noted above, one manufacturer's deflection limit of Span / 480 for header and cross beams appears to be a reasonable starting point for setting deflection limits for a particular project. As is the case with impact factors, the responsible engineer must consider the specific conditions of the lift being planned and adjust this deflection limit accordingly.

6.2.6   Beams of Materials Other Than Steel

The vast majority of gantry system header and cross beams are structural steel, either rolled shapes or fabricated sections. There are some applications, however, where aluminum beams are used. In the author's experience, aluminum beams are used primarily where the contractor wants a beam light enough that workers can carry it and lift it into place atop the gantry legs manually. Thus, the use of aluminum tends to be limited to relatively small extruded beams. Fabricated aluminum beams are typically not used with gantry systems.

The design of aluminum structural members is governed in the United States by the *Specification for Aluminum Structures* (AA 2010). This specification defines design requirements for aluminum members and connections much like those of the AISC specifications for steel. Thus, the use of AA (2010) for the design of header and cross beams is generally appropriate. A few points must be made here, however.

The derivations of the equations presented in this chapter for addressing close proximity concentrated loads (Eqs. 6.9 through 6.11) are based on steel design practice. The use of these equations in conjunction with AA (2010) must be investigated by the responsible engineer prior to implementation. The underhung trolley design provisions adapted from CMAA (2010b) and FEM (1983) are empirical and are based on tests performed on steel beams. Noting that the experimental work considered beams in which the lower flanges did not exhibit yielding, we see that these equations are based on elastic behavior. Thus, the use of these equations for the design of an aluminum beam with an underhung trolley is appropriate.

It is highly unlikely that materials other than steel or aluminum would be used for gantry header or cross beams. However, if such a need arises, the responsible engineer must evaluate the application and apply design methods that will provide a level of structural reliability that is comparable to that provided by the specifications referenced here.

6.3 EVALUATION OF THE GANTRY LEGS

The evaluation of the gantry legs is a very simple task, provided that a few simple rules are followed. This check is as straightforward as comparing the header beam reaction to the rated load of the leg. Nonetheless, the author has seen this simple, yet vitally important, step in the lift engineering process done incorrectly. The appropriate method is outlined here.

The first requirement is to calculate the correct value of the header beam reaction. As noted at the end of Section 6.2.2, the header beam reactions are to be calculated using the actual loads only. Impact factors are not normally used to calculate these reactions. A simple check can be performed to be sure that the correct values are used here. Once the final header beam reactions have been calculated, add up these numbers. If the lift involves side shifting of the suspended load, the reactions used for this check should all be calculated for one position of the load. Next, add up the weight of the load, all rigging, the lift links, the cross beams, the header beams, and any other hardware used in the arrangement that is supported by the gantry legs. These two values, the sum of the header beam reactions and the sum of the weights, must be equal. If they are not, there is an error somewhere in the calculations that must be found and corrected before proceeding.

Now that we know the header beam reactions, what is the rated load to which these reactions are to be compared? Fig. 6.15 is a load chart for the Lift Systems Model 48A gantry system, which is a four-leg system. We can see that the rated load in the first stage at full system pressure is 800 tons (725 tonnes). However, this does not mean that the system is always capable of lifting 800 tons in the first stage. This rated load only holds when the total weight of the lift is shared equally among all four legs.

Many gantry lifts are eccentric, such that the loads to the legs differ front to back, side to side, or both. Thus, we are more interested in the rated load *per leg* than the rated load of the group of legs that make up the system. Continuing with this example, the rated load in the first stage of one leg of a Model 48A gantry system is 800 / 4 = 200 tons (181 tonnes). It is this rated load value that must be compared to the header beam reaction.

A final computation that is often useful is to express the lift as a percentage of the rated load, as is often done for lifts made with mobile cranes. The utilization is calculated as the header beam reaction divided by the gantry leg's rated load. This value must be equal to or less than 100% to be acceptable with respect to the gantry leg capacity. Just as many contractors prefer to limit mobile cranes to working at some reduced capacity, for example, no more than 85% of rated load, some contractors favor similarly limiting the utilization of hydraulic gantry systems. This limit may be a fixed value for all situations, or it may differ depending on the nature of the lift. For example, lifts with travel may have a lower limit than lifts with no horizontal movement and a lift in the top stage of the gantry legs may have a lower limit than a lift that stays in the first stage. Such a practice provides an extra margin that covers the inevitable unknowns that seem to exist on almost every job.

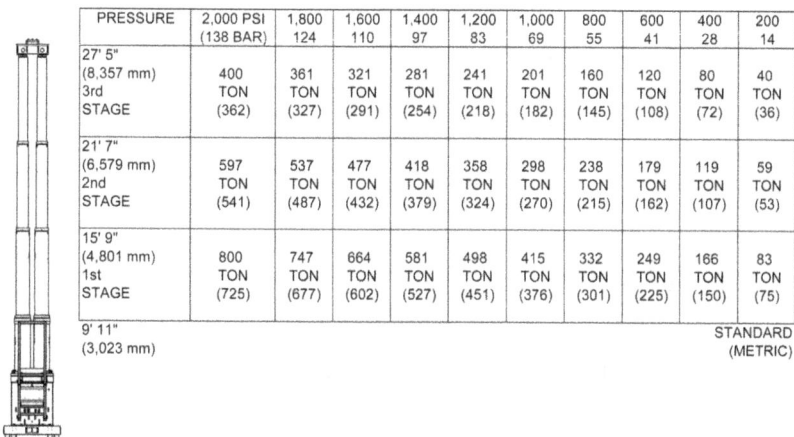

PRESSURE	2,000 PSI	1,800	1,600	1,400	1,200	1,000	800	600	400	200
	(138 BAR)	124	110	97	83	69	55	41	28	14
27' 5" (8,357 mm) 3rd STAGE	400 TON (362)	361 TON (327)	321 TON (291)	281 TON (254)	241 TON (218)	201 TON (182)	160 TON (145)	120 TON (108)	80 TON (72)	40 TON (36)
21' 7" (6,579 mm) 2nd STAGE	597 TON (541)	537 TON (487)	477 TON (432)	418 TON (379)	358 TON (324)	298 TON (270)	238 TON (215)	179 TON (162)	119 TON (107)	59 TON (53)
15' 9" (4,801 mm) 1st STAGE	800 TON (725)	747 TON (677)	664 TON (602)	581 TON (527)	498 TON (451)	415 TON (376)	332 TON (301)	249 TON (225)	166 TON (150)	83 TON (75)
9' 11" (3,023 mm)									STANDARD	(METRIC)

Figure 6.15 Example Gantry System Load Chart

6.4 DESIGN OF TRACK

The design of gantry system track beams follows, to a great extent, the same procedure and methods that were discussed in Section 6.2 for the design of header and cross beams. As with header and cross beams, the design loads include the dead weight of the track assembly, the actual applied vertical loads (which are the gantry leg wheel or roller loads), some specified vertical impact load, and some specified horizontal dynamic load. The required strength of the track beam is again based on its ability to support this combination of specified loads without exceeding appropriate allowable stresses. The required stiffness of the track is based on its ability to deflect no more than some specified amount, normally considering only the actual applied vertical loads. The procedure for designing track beams follows the steps listed below.

- Determine the gantry leg wheels loads, normally assumed to be equal to the sum of the header beam reaction and the weight of the gantry leg divided by the number of wheels;
- Select vertical and horizontal impact factors to account for dynamic loading effects and, if appropriate, gravity load uncertainty;
- Compute the shear, bending, and local compressive stresses in the track beam that result from the combined loads that act during the lift and compare those stresses to the allowable stresses for the beam;
- Calculate the support point reactions of each track beam;
- Evaluate the ability of the underlying surface (soil, structural element, etc.) to safely carry the imposed track beam reactions; and,
- Compute the vertical deflection of the beam due to the actual loads and compare that value to the allowable deflection.

Each of these track design steps is investigated in appropriate detail in the following sections. This discussion is based on the most common gantry leg configurations, in which the leg is fitted with wheels. One gantry system manufacturer produces gantry legs that run on rollers that are similar in configuration to the roller chains used in industrial rollers. The operator's manuals for these gantries include specific instructions and equations for track design. As always, a manufacturer's directions must take precedence.

6.4.1 Design Loads

By the time the lift planner reaches the point of designing or selecting the gantry system track, almost all of the work in determining the acting vertical loads has been completed. The calculation of the loads at each lift link and the positions of those loads is performed as a part of the header and cross beam design. The dead weights of the header and cross beams are also tallied as a part of the design of those beams. One of the important final values that is determined as a part of the header beam design is the maximum header beam reaction, without any impact factors, that is applied to each gantry leg.

The next step in the track design procedure is the selection of dynamic load factors. The SC&RA *Recommended Practices* (SC&RA 2004) suggests that minimum values of 10% vertical impact plus 10% horizontal impact be used for track beam design. Experience has shown that these values produce safe and economical beam designs. As always, the responsible engineer should consider the actual conditions of the lift being planned and revise these impact factors if appropriate.

One question on track beam design loads that is reasonably raised concerns the true effects of lateral loads on the gantry system. A horizontal load created by a swaying load, inertia due to side shifting, or an upending operation will be applied to the system at the top of the leg, as is illustrated in Fig. 6.16. In addition to creating a horizontal reaction at the bottom of the leg, this load will induce an overturning moment on the leg which will, in turn, create a greater reaction on one track beam, as shown in the figure. The most commonly used procedure for designing track beams does not address this potential load imbalance between the two beams in a track assembly.

A second question that has been asked regards the magnitudes of the two dynamic load factors suggested for track beam design. In light of the discussion in Chapter 4 of the dynamic loads that can be expected to act during a lift made with a hydraulic gantry system, values of 10% in both vertical and horizontal directions seem excessive.

The answers to both of these questions come from the same source. The track beam design recommendations of 10% vertical impact plus 10% horizontal impact

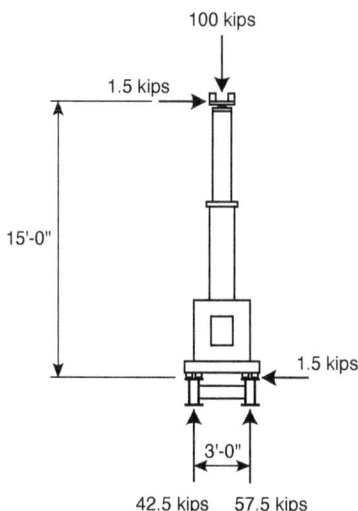

Figure 6.16 Gantry Leg Loads and Reactions

came out of an unpublished study of track beam strength performed by the author in 1994. At that time (as now), the most common method of designing gantry system track beams was based on the assumption that the vertical load from the gantry leg was shared equally among all of the wheels and that the allowable beam strength was based on the provisions of the AISC *Specification* (AISC 1989). The goal of the study was to develop vertical and lateral dynamic load factors for use with this accepted track beam design practice that would produce a required track beam size that would not fail if the fully extended gantry leg toppled laterally. That is, one track beam should be able to carry the full leg load as the leg topples laterally. In this description, "fail" refers to exceeding a strength limit state, not just exceeding allowable stresses.

Using gantry specifications and track beam sizes from one manufacturer's product line, the track beams were evaluated for a range of loads and spans using this standard design method. The result of this first part of the study was a determination of the maximum allowable lifted load for each specified track beam, gantry leg, and span combination using vertical and lateral impact factors of 10%. Both the vertical loads and the horizontal force were distributed equally to all four corners of the gantry leg base.

The actual vertical and lateral bending stresses in the track beams were then computed for the condition at which the supported gantry leg was loaded horizontally with a force great enough to topple the leg. In this case, all vertical and horizontal loading was supported by one track beam only. (Note that this approach is conservative, since the track assembly cross members will distribute the lateral loads to both track beams.) These stresses were then compared to the yield stress of the beam material.

The findings of this study were reasonably consistent. When the gantry leg was loaded to the point of toppling, the peak combined bending stress in the track beam was approximately equal to the yield stress (plus or minus a few percent) for all configurations and beam types checked. This indicates that, although the stresses in such an extreme condition of loading will greatly exceed the normal allowable stresses, the ultimate failure of the system will be toppling of the gantry legs, rather than structural failure of the beams. Thus, it was concluded that the combination of vertical and horizontal impact factors equal to 10% combined with AISC (1989) allowable stresses is a reasonable and practical method of track design. The validity of this approach to track beam design is explored further in Section 6.4.4.

Thus, we see that the generally accepted track beam design method, like the header and cross beam design method discussed in Section 6.2, uses large impact factors in conjunction with a simplified loading model to produce an acceptably safe design. This approach has an advantage beyond the reduction of the complexity of the calculations. A track design method that accounts for the true dynamic loading and load distribution, as illustrated in Fig. 6.16, is dependent upon the gantry leg height. The design method outlined here removes the leg height value from the calculations, thus enabling the engineer to design a track assembly without knowing the specific gantry leg model to be used. This also allows the development of a track load chart from which the gantry system user can see, for example, that a particular track assembly can safely span 20 feet (6 meters) with a total gantry leg load of 100 tons (91 tonnes), provided that the gantry leg wheelbase is at least 54 inches (1,370 mm). This approach has an obvious practical value both for gantry manufacturers and for contractors with a pool of gantry equipment.

6.4.2   Strength Design of Track Beams

Once the design loads have been established, the strength design of track beams generally utilizes the same methods discussed in Sections 6.2.2 and 6.2.3 for header and cross beams. There are a few differences in the design requirements, as discussed here.

Many gantry lifts call for travel of the gantry system along the track beams. Thus, the method of addressing moving loads on a beam (Fig. 6.8) must be applied. There is one significant difference between header beams and track beams, however. The load suspended from a header or cross beam is almost always captured between the beam's supports. A gantry leg can not only move right up to the end of the track beam (Fig. 6.17a), it may straddle a support as the leg moves from one track assembly to the next (Fig. 6.17b).

This movement up to or over a support affects the maximum shear force in the beams and the maximum support reaction where two track sections butt together (reaction R_2 in Fig. 6.17b). The calculation of maximum shear is simply done with the two loads positioned as illustrated in Fig. 6.17a. The load position at which the maximum reaction occurs varies, depending on the values of the loads and the

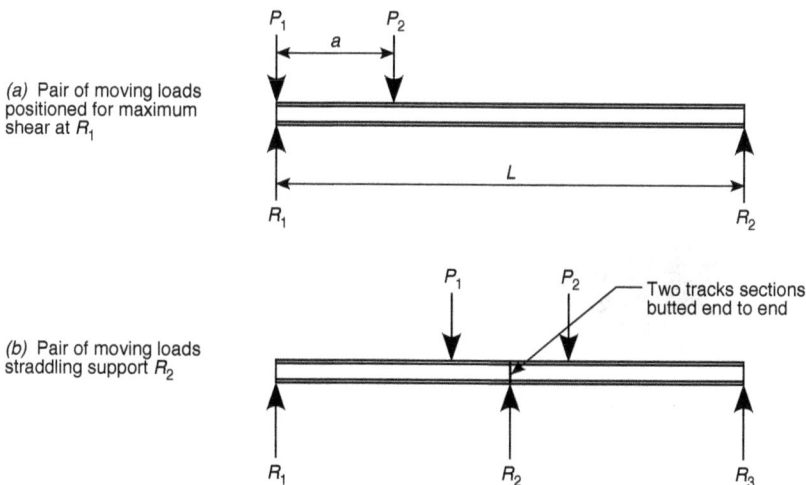

Figure 6.17 Track Beam Loads and Reactions

spans of the adjacent track beams. The more common cases of load positions with respect to maximum reaction are described below.

- If the two loads are equal and the two spans are equal, then the maximum reaction occurs when one load is directly over the support. This value remains constant as the pair of loads traverses the support.
- If the two loads are equal and the spans are different, then the maximum reaction occurs when one load is over the support and the other load is on the longer span.
- If the two loads are different, as occurs when two differently loaded legs are close to one another, and the two spans are equal, then the maximum reaction occurs when the greater load is directly over the support.
- If the two loads are different and the two spans are different, the maximum reaction will occur when one load is over the support. However, which load is over the support at this point depends on the difference between the loads and the difference between the spans, so the various possible positions of the loads must be checked.
- Last, when more than two loads must be considered, as occurs when two gantry legs are traveling along two relatively long track spans, these generalities do not apply and all of the various possible load positions must be analyzed.

Pairs of concentrated loads in close proximity to one another occur in track design when evaluating track for use with some of the larger gantry models that use wheel assemblies with two axles (Fig. 6.18). The methods described in Section 6.2.3 can be applied to track design to handle this condition. As noted in that discussion, the length of bearing of the concentrated load l_b used in the equations is taken as zero for a wheel or roller.

Figure 6.18 Closely Spaced Gantry Leg Wheels *(David Duerr, P.E.)*

One must note that the equations in Section 6.2.3 only address concentrated loads along the span of the beam, not at the ends, and that the design provisions in AISC (1989) and AISC (2010b) both have different requirements for concentrated loads at or near the end of a beam. For practical reasons, this condition is not normally an issue in gantry track design. Virtually all gantry track beams have plates across the width of the beams at both ends that provide a means to connect the track beams end to end (Fig. 6.19). These plates also function as web stiffeners with respect to supporting concentrated loads.

There is one more condition of a pair of concentrated loads that was not addressed for header beam design, but that does apply to track beam design. That is web compression buckling, a limit state that occurs when loads in line with one another on opposite sides of a beam place the web in compression. This condition occurs in a gantry track beam when a pair of wheels, such as the wheel assembly seen in Fig. 6.18, passes over a support at an unstiffened point along the length of the beam.

Studies performed by the author, again using the finite element software BASP, show that as long as the spacing of the two loads s is equal to or greater than twice the clear web height between the beam flanges h, then the loads act independently and the AISC (1989) or AISC (2010b) provisions can be applied without modification. If the two loads are more closely spaced, then the allowable compression load R calculated using the AISC equations must be reduced by the correction factor c given by Eq. 6.43.

$$c = 0.5 + \frac{s}{4h} \leq 1.0 \qquad (6.43)$$

Since we are only concerned here with the loads acting to compress the web from both flanges, the applied load that is compared to the corrected AISC

Figure 6.19 Track Beam End Detail *(David Duerr, P.E.)*

allowable load for web compression buckling is the gantry leg wheel load, not the beam support reaction.

Some gantry leg models run on roller chains, rather than wheels. Due to the close spacing of the individual rollers, the various web strength limit states can be checked by treating the load as being applied as uniformly distributed over the length of the roller chain. That is, the length of bearing l_b is taken as the overall length of the roller chain that is in contact with the top flange of the track beam. This approach presumes that the track beam support and stiffness are such that the beam truly does support the roller chain in this manner. However, two of these web strength checks will require further work on the part of the lift engineer.

Examination of the equations in the AISC specifications for the limit states of web sidesway buckling (which is only applicable to single-web beams) and web compression buckling reveal that these equations do not contain the term l_b for the length of bearing. The derivations of these equations are based on the assumption that the compressive loads are applied as point loads (Summers and Yura 1982 for web sidesway buckling) or over a very small length of bearing (Chen and Newlin 1973 for web compression buckling). The long length of bearing of a roller chain on the top flange of a track beam creates a significantly different stress distribution in the web compared to that resulting from a point load, so direct application of the AISC (2010b) equations will produce questionable, albeit typically very conservative, results. It is incumbent upon the responsible engineer to investigate the research that underlies these equations and to revise the calculation method accordingly to arrive at a technically sound result. (Although the author has not pursued this in depth, it appears that comparison of the experimental results reported in Chen and Newlin 1973 to the theoretical solutions derived in Chapter 9 of Timoshenko and Gere 1961 may be useful.)

6.4.3   Track Beam Deflections

As is the case with header beams and cross beams, the need to limit the deflections of track beams is likewise not agreed upon within the industry. One gantry system manufacturer recommends that the mid-span deflection of track beams be limited to 1/4 inch in 20 feet (5.2 mm in 5 meters), which is equal to Span / 960. Others experienced in the design and use of hydraulic gantry systems do not feel that specific track beam deflection limits are needed. SC&RA (2004) again only directs the lift planner to the gantry system manufacturer for guidance on deflection limits. One issue that is agreed upon for track beams as for header beams is that when deflections are calculated, the calculations are performed using only the gravity loads supported by the beam.

Limiting track beam deflections is more important than limiting header and cross beam deflections. As was discussed in Section 6.2.5 with respect to side-shifting a load, there is always a concern about controlling the movement of the gantry legs, themselves, along the track. Excessive deflections of the track beams creates the same "uphill" / "downhill" problems discussed previously with respect to header beams and cross beams.

Perhaps of greater concern is the effect of track beam stiffness on the stability of the gantry system. The stability of the system depends, to a great extent, on the both strength and the firmness of the supports. Stringent deflection limits for the design of track beams will help assure a stiff support for the legs which will, in turn, limit the problems that can arise from unequal deflection of the track beams. This value can be illustrated by means of an example.

EXAMPLE 6-1

Consider the gantry leg illustrated in Fig. 6.16. In the static condition, the load to each track beam is 50 kips (22.7 tonnes). The track beam span L is 10 feet (3.05 meters) and assume that the beam design is such that the mid-span deflection in this condition is L / 960 = 0.125 inch (3.2 mm).

When the indicated horizontal load is applied, the beam loads change to 57.5 kips (26.1 tonnes) on the heavy side and 42.5 kips (19.3 tonnes) on the light side. Thus, the beam deflection on the heavy side will be 0.125 / 50.0 x 57.5 = 0.144 inch (3.65 mm) and the beam deflection on the light side will be 0.125 / 50.0 x 42.5 = 0.106 inch (2.70 mm). The difference in deflection is thus 0.144 - 0.106 = 0.038 inch (0.97 mm). Given the track gauge of 36 inches (914 mm), the slope across the track beam gauge that results from this differential deflection is only 0.038 / 36 x 100% = 0.11%.

This differential deflection can also be compared to the common track setup tolerance of 0.125 inch (3.2 mm) across the track gauge. The difference is only 0.038 / 0.125 * 100% = 30% of the track setup tolerance.

We can see from this example that limiting the deflections of track beams under the static load case, which is the case for which track beam deflections typically are calculated, will limit the differential deflections of the beams when the gantry system is subjected to a lateral horizontal load. This serves to maintain the levelness of the gantry leg support and, in turn, the stability of the system.

In summary, it is suggested that the mid-span deflections be calculated for track beams as a part of the engineering of a gantry lift. The previously stated deflection limit of 1/4 inch in 20 feet (Span / 960) for track beams is suggested for general use. This value should be treated as a guideline, however, and not as an absolute requirement. The engineer responsible for the planning of the lift must decide if this limit is appropriate for the particular conditions at hand and should use a different limit, either larger or smaller, if warranted. Of course, a directive from the gantry system manufacturer should be followed.

It is important to note that the supports under the track beams must similarly be stiff and possess adequate strength to support the maximum track reactions. The supports should also support both beams of the track assembly similarly; i.e., the spans of the two beams of a track assembly should be equal and the support stiffnesses under each beam should also be as close to equal as possible. A properly designed track beam bearing on poor quality cribbing will not give adequate support to the gantry system. Rather, the track must be viewed as a system consisting of the track beams, any supporting structural elements, such as stands, cribbing or mats, and the underlying base, which may be a building floor or other structure, a slab on grade, or soil. Only when all of the elements of this system are sound will the desired level of support be provided. Consideration of the overall track system is a classic example of a chain being only as strong as its weakest link. The engineering and planning must carefully examine every one of the links in this chain.

6.4.4   Evaluation of the Track Beam Strength Design Method

Sections 6.4.1 and 6.4.2 discuss the loads and strength design methods to be used for the design of gantry system track. As is discussed in Section 6.4.1, relatively large impact factors are commonly used to allow application of a somewhat simplified strength design method.

One may reasonably question how well this approach to track design works. We will examine this question by means of two examples. Example 6-2 looks at the effects on the track beams of the concentrated loads from the wheels. Example 6-3 considers overall bending of the beams.

EXAMPLE 6-2

Gantry leg specifications:
Rated load = 110 tons (100 tonnes) at a height of 25 feet (7.6 meters)
Rated load = 80 tons (72.6 tonnes) at a height of 32 feet (9.8 meters)
Dead weight = 20,000 pounds (9,072 kg)
Track gauge = 36 inches (914 mm)
Two wheels per corner; eight wheels total

Case 1a - Lift at maximum rated load - standard design method
Gantry leg load to track = 1.10 x (110 x 2,000 + 20,000) = 264,000 pounds
Wheel load = 264,000 / 8 = 33,000 pounds

Case 1b - Lift at maximum rated load - account for actual load application
Gantry leg load to track = 110 x 2,000 + 20,000 = 240,000 pounds
Lateral load at top of leg = 0.015 x 110 x 2000 = 3,300 pounds
Track transverse slope due to 1/8" out-of-level = sin⁻¹ (0.125 / 36) = 0.20 degree
Lateral load effect due to out-of-level track = 220,000 sin (0.20°) = 768 pounds
Total acting lateral load = 3,300 + 768 = 4,068 pounds
Load to heavy side track beam = 240,000 / 2 + 4,068 x (25 x 12) / 36 = 153,900 pounds
Wheel load = 153,900 / 4 = 38,475 pounds = 1.166 times design wheel load.

Case 2a - Lift at rated load at extended height - standard design method
Gantry leg load to track = 1.10 x (80 x 2,000 + 20,000) = 198,000 pounds

Wheel load = 198,000 / 8 = 24,750 pounds

Case 2b - Lift at rated load at extended height - account for actual load application
Gantry leg load to track = 80 x 2,000 + 20,000 = 180,000 pounds
Lateral load at top of leg = 0.015 x 80 x 2000 = 2,400 pounds
Track transverse slope due to 1/8" out-of-level = $\sin^{-1}$ (0.125 / 36) = 0.20 degree
Lateral load effect due to out-of-level track = 160,000 sin (0.20°) = 559 pounds
Total acting lateral load = 2,400 + 559 = 2,959 pounds
Load to heavy side track beam = 180,000 / 2 + 2,959 x (32 x 12) / 36 = 121,563 pounds
Wheel load = 121,563 / 4 = 30,391 pounds = 1.228 times design wheel load.

We see in the two cases in this example that the actual maximum wheel load due to the vertical load, the track transverse slope, and a transverse horizontal load equal to 1.5% of the lifted load will be greater than the wheel load calculated using the standard track beam design method. The greater difference in the example is 22.8%. Application of this analysis to other gantry leg models yields similar results, with the percentage difference not exceeding 25% for the medium and high capacity gantry legs.

The use of the AISC allowable strength equations provides a design factor of 1.50 or greater for the concentrated load limit states (web local yielding, web local buckling, web sidesway buckling when applicable, and web compression buckling). If a track beam was proportioned such that one or more of these concentrated load limit states was approaching its allowable value using the standard beam design method, we can see that the true wheel loads would create an overstress relative to the provisions of the design specification, but would not be approaching an actual failure. This is consistent with the thinking that underlies the original adoption of the 10% vertical impact + 10% lateral impact + AISC allowable stresses gantry track beam design basis in the 1990s.

An important variable for the examination of track beam bending is the ratio of major axis bending strength to minor axis bending strength. This ratio varies significantly among the different proportions of track beams commonly used, ranging from about 2.00 to 4.00 for fabricated box beams and wide flange beams built up with side plates to form box sections.

EXAMPLE 6-3
Gantry leg specifications:
Rated load = 110 tons (100 tonnes) at a height of 25 feet (7.6 meters)
Dead weight = 20,000 pounds (9,072 kg)
Track gauge = 36 inches (914 mm)
Wheelbase = 60 inches (1,524 mm)
Two wheels per corner; eight wheels total

Track specifications:
Span = 15 feet (4.57 meters)

Major axis bending capacity = 5,000 kip-inches (565 kN-m)
Ratio of major axis bending strength to minor axis bending strength = 2.00

Case 1 - Standard design method
Corner vertical load = 1.10 x (110 x 2,000 + 20,000) / 4 = 66,000 pounds
Corner lateral load = 0.10 x (110 x 2,000 + 20,000) / 4 = 6,000 pounds
Major axis live load moment = 4,125 kip-inches (see Section 6.2.2 for the calculation method)
Minor axis live load moment = 375 kip-inches
Bending interaction ratio = 4,125 / 5,000 + 375 / (5,000 / 2.00)
Bending interaction ratio = 0.98

Case 2 - Account for actual load application
Gantry load to track = 110 x 2,000 + 20,000) = 240,000 pounds
Total acting lateral load = 4,068 pounds (from Example 6-2, Case 1b)
Load to heavy side track beam = 240,000 / 2 + 4,068 x (25 x 12) / 36 = 153,900 pounds
Corner vertical load = 153,900 / 2 = 76,950 pounds
Corner lateral load = 4,068 / 4 = 1,017 pounds
Major axis live load moment = 4,809 kip-inches
Minor axis live load moment = 64 kip-inches
Bending interaction ratio = 4,809 / 5,000 + 64 / (5,000 / 2.00)
Bending interaction ratio = 0.99

We see that, in this case, the bending interaction ratios computed using the two methods are almost identical. We can also see that as the ratio of major axis bending strength to minor axis bending strength increases above the lower bound value of about 2.00, the standard track beam design method becomes increasingly conservative. Thus, taken as a whole, the standard method (10% vertical impact factor, 10% lateral impact factor, loads applied through the centroidal axes of the beam, strength based on a nominal design factor of 1.67) achieves the original goal of providing a safe, practical, and efficient method of designing track beams for hydraulic gantry systems.

Noticeably absent from this discussion of track beam strength design is any mention of torsion. This is so for one simple reason: except in the most unusual of cases, torsion is negligible in the design of gantry system track beams.

As seen in Example 6-3, the actual transverse load applied to a track beam is relatively small. Including the effect of an out-of-level track setup, the total transverse load is less than 2% of the gantry leg corner load. This load is applied at the top surface of the beam, so the torsion applied at one corner of the gantry leg is only $F_{tr}(d / 2)$, where F_{tr} is the transverse load at one corner of the leg and d is the depth of the track beam. Torsion can also occur due to an off-center vertical load. This torsion, too, should be trivial if the track assembly is detailed properly with a suitably sized wheel guide bar. Track beams for all but the very lightest gantry systems are box shapes, which have high torsional strength and stiffness.

Together, the combination of a small applied torsional moment and high torsional strength results in a very low shear stress due to torsion in the beams. Thus, these effects are generally trivial and torsion calculations are rarely required as a part of gantry track beam design.

6.4.5   Track Assembly Appurtenance Design

The analysis of the track beams in the overall track assembly is typically all that is required as a part of the engineering for the planning of a particular lift. There is, however, more to the assembly than just the beams and these other components require attention when designing a track assembly from scratch.

A basic track assembly of the style most commonly used is illustrated in Fig. 6.20. The additional components that must be detailed as part of a complete track assembly design are the cross members, the end connection plates, the wheel guide bar, and (optionally) the propel cylinder anchor. The design of these components is not addressed in any of the standards or guides published to date, so the responsible engineer must establish criteria. A few comments are offered here to provide guidance.

The cross members serve to hold the two track beams at the required gauge and to prevent the beams from rotating about their longitudinal axes under the effects of transverse loading or off-center vertical loading. Thus, an appropriate design load for a cross member will include an end moment equal to the maximum torsion that may be expected to be applied to a track beam. If the track assembly is to be used with a gantry system that utilizes external propel cylinders, the cylinder reaction will apply both vertical and horizontal load components to the cross members. These loads will develop major axis and minor axis moments in the cross members. The potential magnitudes of these loads are discussed in

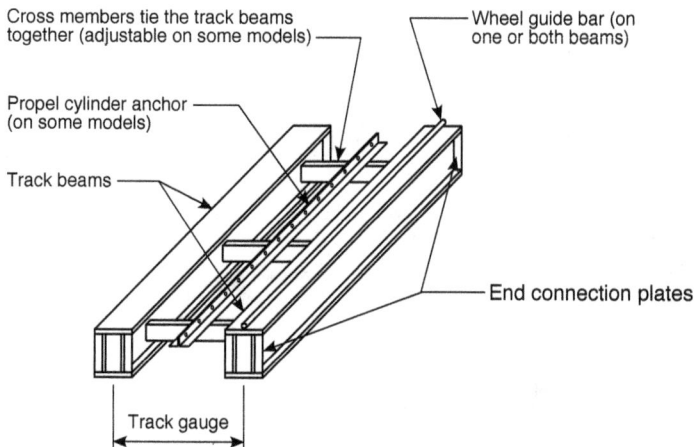

Figure 6.20 Gantry System Track Assembly

Section 6.7. The cross members may be fixed in length or may be adjustable to allow using the track assembly with gantry legs of different gauges.

The end connection plates on most track assemblies are designed as simple shear connections. The most important function of the end connection detail is to assure that the two track assemblies that are connected together are accurately aligned vertically and horizontally. Precise alignment allows the gantry leg to roll from one track beam to the next with minimal resistance. These track beam joints should be fully supported so that the applied wheel loads are transferred directly through the beam to the support. In this manner, the actual shear force carried by the end connection will always be minimal. However, if a particular track setup will impose a large shear force at the connection, the appropriate analyses of the connectors (bolts or pins), the end connection plates, and the welding between the plates and the beams must be performed.

The wheel guide bar does exactly what the name implies. It provides restraint, or guidance, to keep the wheels or rollers centered on the track beam. The detailing of the wheel guide bar is more a function of geometry than strength. The bar must be large enough that the wheels won't "walk up" onto the bar during travel, but it must be small enough to fit in the available space. A round bar stitch-welded to the top flange of the track beam is the most common style of guide bar in use today.

A close-up view of a wheel guide bar is seen in Fig. 6.21. One clearance dimension is seen in the photograph. The diameter of the guide bar must be smaller than the horizontal clear space between the wheels. There is also a vertical clearance that must be considered, as well. Many gantry models have little height between the underside of the base weldment and the top surface of the track. This height must be checked to be sure that the underside of the gantry leg won't rub on the top of the wheel guide bar. Last, the welding of the guide bar to the track beam top flange must be detailed such that the welds do not encroach on the track surface that the wheels contact.

Figure 6.21 Wheel Guide Bar *(David Duerr, P.E.)*

The last component to be discussed here is the propel cylinder anchor. This longitudinal member provides a reaction point against which the propel cylinder will push or pull. As noted in the discussion on cross members, the propel cylinder reaction will have both vertical and horizontal components. These forces will develop an axial load and a bending moment in the propel cylinder anchor and a shear force in the connections between the anchor member and the cross members. Propel cylinder loads are discussed in Section 6.7. Design of the propel cylinder anchor and the cross members with respect to propel loads may be performed with either the design propel system forces or with the maximum load that may be imparted by the propel cylinder. This latter method, while unquestionably conservative, assures that the track assembly components won't be the "weak link" in the design.

The use of the external propel cylinders creates an additional demand on the connections between track assemblies. It is possible for the gantry legs to be on one track assembly and the propel cylinders to be anchored to the adjacent assemblies. This will place the connections between track assemblies in tension when the propel cylinders are pushing the gantry legs. Thus, the end connection plates and their associated connectors and welds must also be designed to carry this tension load in combination with any other design loads. Again, the conservative approach is to size these parts for the full load that may be imposed by the propel cylinders.

6.4.6   Track Beam Support

When the track beams must span a significant distance, the calculation of the end reactions is a standard part of the check of the beam strength. This is discussed in Section 6.4.2. A more common track support arrangement calls for a series of timbers placed perpendicular to the track beams and spaced relatively closely, as illustrated in Fig. 6.22. The appearance of this support arrangement is much like that of railroad tracks on ties.

This track support arrangement may be used on bare soil, on paving, and on more substantial structures, such as an equipment foundation. Consequently, the demands placed on the cribbing timbers varies from simply providing part of a means of leveling the track beams to providing load spreading to a subsurface of limited load bearing capacity.

The load imposed on each timber depends on the load on the gantry leg, the spacing of the timbers, the wheelbase of the gantry leg, and the spacing between legs when two legs are on one track section. The load distribution among the timbers also depends on the stiffness of the surface on which the timbers bear and the bending stiffness of the track beams. As an example, consider the arrangement illustrated in Fig. 6.22. Here we have two gantry legs spaced 12'-0" center-to-center rolling on a track section that is supported by timbers spaced 5'-0" center-to-center. The support reactions to each timber due to the loads imposed by the

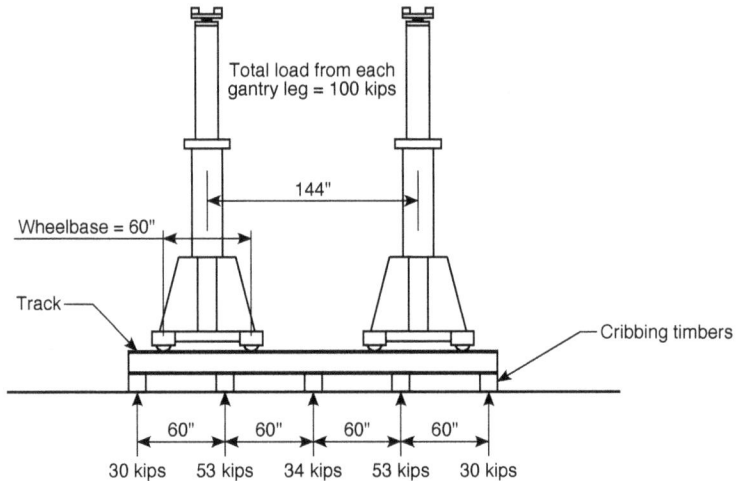

Figure 6.22 Track Support Cribbing Reactions Under Two Track Beams

two gantry legs are shown, based on a very stiff supporting surface. As one would expect, these support reactions will change as the pair of gantry legs travels along the track beams.

These support calculations are the repeated for the same setup with one change. In Fig. 6.23, the two gantry legs are only 7'-6" center-to-center. We see that the maximum support reaction for this arrangement is significantly greater than that seen in Fig. 6.22. And again, these values will change as the pair of gantry legs move along the track.

If these support reactions were calculated based on relatively soft supports, as would occur if the timbers were placed on bare soil, the numbers would tend to even out due to the stiff track beams serving to spread the load. In general, stiff track beams and soft supports will move toward a greater, more uniform sharing of the load among the timbers. A very stiff supporting surface will give results more in line with those seen in Figs. 6.22 and 6.23.

Due to the many variables that affect the load distribution (as listed on the previous page), a reasonably accurate calculation of the track support reactions requires a structural analysis of the beams that takes into account all of the layout dimensions, the track beam properties, and the nature of the supporting surface. Experience tells us that the information needed to accurately assess the stiffness of the supporting surface is often unavailable, so a more practical approach is needed. The following guide can be used when the performance of a more detailed analysis is not practical. This guide applies when the support spacing is equal to or less than the gantry leg wheelbase.

- When just one gantry leg is on the track section, the maximum load to one support under one track beam may be as much as 40% greater than the load at one corner of the leg as the leg straddles that support point.

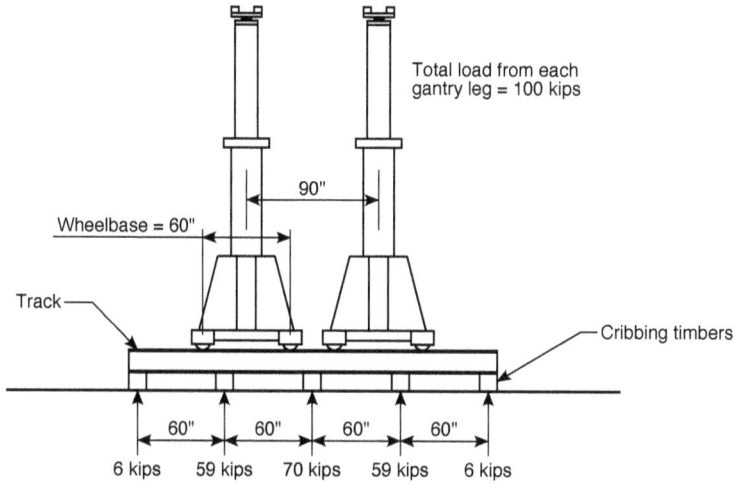

Figure 6.23 Track Support Cribbing Reactions Under Two Track Beams

- When two gantry legs are on the track section and the legs are spread apart (similar to Fig. 6.22), the maximum load to one support under one track beam will be similar to the case of a single gantry leg on the track. That is, the support load may be as much as 40% greater than the leg corner load.
- When two gantry legs are on the track section, the legs are close together, and the gantry leg loads are close to equal (similar to Fig. 6.23), the maximum load to one support under one track beam may be as much as 50% greater than the gantry leg corner load.
- When the cribbing timber spacing is significantly less than the gantry leg wheelbase and the supporting surface is relatively soft, such as bare soil, the maximum load to one support under one track beam may be less than the gantry leg corner load. Assuming that the load from the gantry leg is shared by the minimum number of timbers that may be under the leg as the leg moves may be a reasonable assumption.

As can be seen in these descriptions, a certain amount of judgment must be exercised to assess track support reactions when the information needed to perform a detailed analysis is not available. Given the importance of firm support to the stability of a gantry system, conservative assumptions are usually best.

Once the track beam reactions have been calculated, or at least estimated with acceptable accuracy, the adequacy of the components below the track can be evaluated. Let us use the track support values shown in Fig. 6.23 to illustrate the most important items to check.

The maximum reaction is 70,000 pounds from two track beams, or 35,000 pounds each. The cribbing is drawn as 12" x 12" timbers and we will assume that the track beam width is 10". Thus, the bearing stress between a track beam and a cribbing timber is 35,000 / (12 x 10) = 292 psi. The allowable bearing stress perpendicular to the grain of common hardwoods is typically in the range of 700 psi to 800 psi (AWC 2012), so we can see that bearing to the timbers is acceptable. However, the allowable compressive stress of plywood is typically about 160 psi (APA 1998). This tells us that, for this setup and loading, plywood is not an acceptable shim material.

The next step is the calculation of the bearing pressure between the timber and the underlying surface. Unless the timbers are unusually long, this is commonly assumed to be equal to the total supported load divided by the footprint area of the cribbing timber. If the timbers are, for example, 5'-0" long, then the bearing pressure is 70,000 / (1 x 5) = 14,000 psf. This pressure is then compared to the allowable bearing pressure for the surface.

Last is the evaluation of the strength of the cribbing timbers. This is a simple beam design problem based on the length of the timber, the spacing of the track beams, and the bearing pressure between the timber and the surface below. The strength design can be based on the provisions of AWC (2012) in the United States or a comparable construction design standard in other countries. The allowable stresses used in this evaluation must be consistent with the type of wood and the condition of the timbers. If the timbers are fairly long, a deflection analysis must be performed to determine the true effective length of bearing between the timbers and the underlying surface (Duerr 2010).

6.5   OTHER BEAM DESIGN ISSUES

One area of concern in the design of header beams and cross beams is that of potential problems in the use of very heavy wide flange shapes as beams. Wide flange shapes such as W12x230 and heavier and W14x283 and heavier [$t_f > 2$" (51 mm)] are commonly used as header and cross beams. Shapes of this size, commonly referred to as jumbo beams, are generally produced by the steel mills for use as columns and other compression members. When subjected to tension stresses, as happens in beams, cracking can occur under certain conditions.

An extensive study of this problem was performed a number of years ago (Fisher and Pense 1987) as a result of some failures of trusses fabricated with jumbo beams. An important conclusion of this study is that these cracks are always associated with groove welded splices. Rolled members that have not been spliced by groove welding and that have not been flame cut with access holes or the like did not exhibit the development of cracks that can lead to a tensile failure.

Based on the results of this study, one may reasonably conclude that the use of as-rolled jumbo wide flange shapes as header beams and cross beams is an acceptable practice. Given the long history of the use of jumbo beams with hydraulic gantries, this conclusion is not surprising. Of course, splicing two jumbo beam segments together by welding should be avoided.

The second subject to be addressed here is that of fabrication tolerances. This is particularly important for gantry system track beams. The gantry leg wheels or rollers must bear evenly on the top flange of each track beam. This tells us that the track beams must be nominally free of twist and warping of the top flange over their full length. The beams must also have little or no sweep and the two beams of each assembly must be parallel. This is particularly important if wheel guide bars are installed on both beams.

The applicable industry standards and guides do not address track fabrication tolerances and the tolerances that apply to convention steel fabrication, such as those given in AISC (2010a), may not be precise enough for the manufacture of gantry track assemblies. Due to the differences in gantry leg support (one, two, or four wheels per corner, or a roller chain along the base), the definition of a "one size fits all" set of track beam fabrication tolerances is not practical. Thus, a contractor interested in fabricating its own track must turn to the gantry manufacturer for guidance on track fabrication tolerances that are compatible with the gantry system to be used.

The last subject that we will address within the realm of header beam and track beam design is the use of load charts. One gantry manufacturer has published header beam load charts that are integrated into the gantry system capacity chart. Rated loads for the overall two-leg system are shown on these charts, some of which are limited by gantry leg capacity and some of which are limited by beam strength. While these charts can be useful, a word of caution is in order. If you choose to use a header beam or track beam load chart, be sure that you understand what design factors were used to determine the stated rated loads. As has been discussed here, varying lift conditions may make advisable larger impact factors, smaller deflection limits, or other deviations from standard practice.

One practical approach to beam design or selection is to use a load chart for the initial header beam or track selection and then perform complete detailed calculations for the final lift planning. This process offers efficiency without compromising the safety of the final lift plan.

6.6 DESIGN OF LIFT LINKS

The design and fabrication of lift links are most commonly tasks performed by the gantry system manufacturers. However, the author has noted that gantry system users are increasingly opting to design their own lift links, both for routine use and to fulfill the needs of unique applications. Due to the lack of guidance in the standards and in industry publications, the author has developed his own lift link design method, the derivation of which is presented here.

In the 1970s and 1980s, engineers typically designed gantry system lift links using elastic analysis methods and allowable stresses. As noted on page 141, the use of AISC (1989) allowable stresses with their nominal design factor of 1.67 was a common and accepted practice. This approach has two shortcomings. First, an elastic analysis methodology for lift link design is not the most practical with respect to understanding the true strength of the link. Lift link usage is rarely, if ever, frequent enough that fatigue is a design issue and the proportions of lift links are such that service load deflections, other than deformation of a hole due to bearing of a pin, are not of concern. Thus, design based on failure limit states, such as critical sections reaching full plasticity, more realistically defines the useful strength of the links. Second, a design factor of 1.67 appears to be low in light of the loading that may be imposed due to cross cornering. A design factor in excess of 2.00 is more appropriate.

Some insight into the service load stress distribution in a lift link can be seen by performing a finite element analysis (FEA) of a link. The results of a linear FEA of a typical plate lift link is shown in Fig. 6.24. Areas of higher stress are more lightly shaded. We can see that the critical areas in the link are in the top element above the header beam (bending, shear, and bearing), at the corners of the rectangular beam opening (bending and tension), and in the region below the shackle pin hole (bending, shear, and bearing). The high tensile stresses in the corners of the beam opening are "hot spot" stresses created by the geometry of the part. The spread of these areas of high stress is blunted as local yielding occurs. As a result, these hot spots are not a concern with respect to the failure load of the link. These corners are typically fabricated with a small radius, perhaps 1/4 inch (6 mm), to minimize crack initiation.

Critical planes that must be analyzed are illustrated in Fig. 6.25. To perform these analyses, we must understand how to calculate the limit state strength of each section under the combined effects of tension, shear, and bending. The following calculation methods and equations are based on the performance of ductile steel, the behavior of which can be approximated as an elastic-perfectly plastic material. By comparison of the drawing in Fig. 6.25 to the stress patterns seen in Fig. 6.24, we can surmise that most of these critical planes will most likely not control the design. However, it is prudent to check all critical sections so as not to miss the "weakest link" in a particular design.

The limit state of yielding in tension T_y of a rectangular section is simply the cross sectional area A multiplied by the yield stress F_y of the material. T_y is expressed as Eq. 6.44.

Figure 6.24 Lift Link Linear Stress Analysis Results

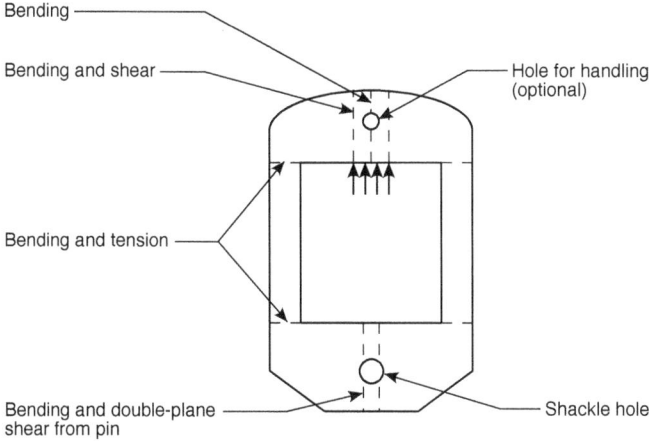

Figure 6.25 Critical Sections in a Basic Plate Lift Link

$$T_y = F_y A \qquad (6.44)$$

The limit state of yielding in shear V_y of a rectangular section is calculated in a similar manner. That is, the cross sectional area is multiplied by the shear yield stress F_{ys}. Current structural steel design practice in the United States defines the shear yield stress of steel in terms of the tensile yield stress and the Huber-vonMises Energy of Distortion Theory. This provides the basis for Eq. 6.45.

$$V_y = F_{ys} A = \frac{F_y}{\sqrt{3}} A \qquad (6.45)$$

The limit state of a plastic moment M_p of a rectangular section is the plastic modulus Z of the section multiplied by the yield stress (Eq. 6.46).

$$M_p = F_y Z \qquad (6.46)$$

The three limit states expressed in Eqs. 6.44, 6.45, and 6.46 only hold when each type of stress acts alone. When any two or all three types of stress act on a section, then an interaction equation must be applied. An appropriate solution is derived in Neal (1961), shown here as Eq. 6.47.

$$\frac{M}{M_p} + \left(\frac{T}{T_y}\right)^2 + \frac{\left(V/V_y\right)^4}{1 - \left(T/T_y\right)^2} = 1 \qquad (6.47)$$

where
M = moment applied to the section;
T = tensile force applied to the section;

V =   shear force applied to the section; and,
             all other terms are as previously defined.

The next issue to be addressed is that of the pin hole through which the load is applied to the lift link. Methods for the analysis of pin-connected plates are derived in Duerr (2006). Given the proportions of lift links as are being considered here, the only strength limit state of concern is that of double plane shear.

The shear failure planes are shown in Fig. 6.26. Experimental research has shown that the angle ϕ that locates the shear planes is about 55° for a snug-fit pin. For the condition in which the pin is smaller than the hole, ϕ is calculated using Eq. 6.48.

$$\phi = 55\frac{D_p}{D_h} \tag{6.48}$$

The shear strength of the material below the pin hole is based on the length of the shear planes Z. The shear plane length is a function of the distance from the center of the hole to the edge of the plate, the pin diameter, and the locating angle (Eq. 6.49).

$$Z = a + \frac{D_p}{2}\left(1 - \cos\phi\right) \tag{6.49}$$

Eq. 6.49 is based on an edge of the plate that is straight and perpendicular to the line of action of the load applied to the lift link. Deviations from this model, such as a curved plate edge, must be accounted for in determining the shear plane length.

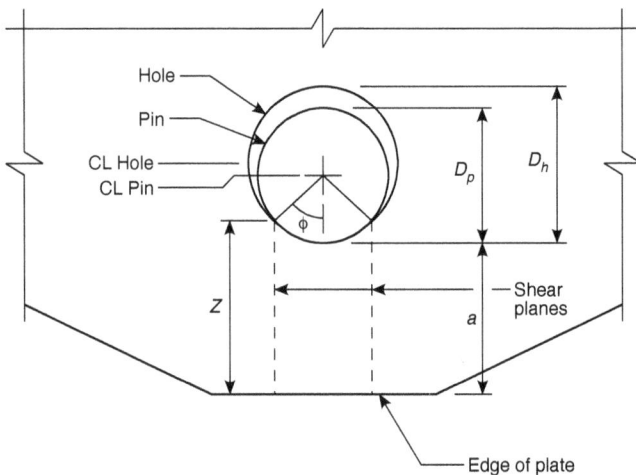

Figure 6.26 Shear Failure Planes Below the Pin Hole

With the length of the shear plane Z determined and noting the thickness of the plate as t, we can calculate the limit state of shear yielding on each shear plane in the region below the pin hole using Eq. 6.50.

$$V_h = \frac{F_y}{\sqrt{3}} Zt = \frac{F_y}{\sqrt{3}} A_v \tag{6.50}$$

We will also examine the bearing stress between the pin and the hole. This check, however, is not a strength concern. Rather, the bearing stress of a pin is typically limited to control excessive local deformation of the plate. Pin bearing stress is commonly calculated (AISC 1989, AISC 2010b, ASME 2012) as the load applied by the pin divided by the projected area of the pin. The allowable bearing stress given in AISC (1989) and AISC (2010b) both equate to 0.90 F_y with respect to the service load. The allowable pin bearing stress in ASME (2012) is 1.25 F_y / N_d, where N_d is the nominal design factor. If we let $N_d = 1.67$, which is the nominal design factor upon which the allowable stresses of AISC (1989) are based, then the allowable bearing stress is 0.75 F_y. This lower value is used in ASME (2012) to further limit deformation of the plate and to reduce bending in shackle pins.

The limit state of bearing can be determined by setting the design factor to 1.00. By doing this, we find the limit state bearing stress is 1.67 x 0.90 = 1.50 F_y in AISC (1989) and 1.25 F_y in ASME (2012). Using the allowable bearing stress value that underlies the AISC provisions, the pin load based on the limit state of bearing is given by Eq. 6.51. This value must be divided by an appropriate design factor to arrive at an allowable service load. (The greater limit state bearing stress is reasonable for lift link design when one considers the design factor of 2.25 discussed at the end of this section. This results in a practical service load bearing stress.)

$$P_b = 1.50 F_y D_p t \tag{6.51}$$

The last issue to be addressed is the support of the lift link. There are three generalized conditions that may apply. These are support on a single-web beam, support on a double-web (box) beam, or support on another type of surface, such as a side shift device. Support on a single-web beam is most severe.

The support of a lift link on a single-web beam is illustrated in Fig. 6.27. The basic question to be considered here is: What is the effective width of bearing between the upper segment of the lift link and the top surface of the beam? To provide insight into the interaction between the flange and the lift link, we must consider the relative stiffnesses of the lift link and the beam flange. If the bending stiffness of the beam flange is significantly greater than that of the top segment of the lift link, then the flange would provide support to the link across much, if not all, of its width. On the other hand, if the bending stiffness of the flange is small compared to that of the lift link, then the support provided to the link by the beam would be concentrated along a small width centered above the web. Since the

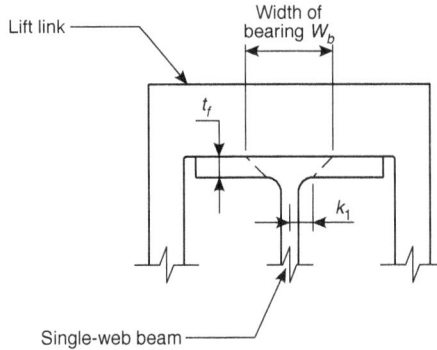

Figure 6.27 Lift Link Support on a Singe-Web Beam

design of a lift link is rarely made based on its use with only one size beam, a lower bound width of bearing must be used.

A width of bearing W_b equal to $2 k_1 + 2 t_f$ is illustrated in Fig. 6.27. That is, hard bearing is assumed to be provided by the beam over a width on either side of the beam centerline equal to the width of web-to-flange fillet (dimension k_1) plus a spread on a 1 : 1 slope through the thickness of the flange. Stiffness analyses of a range of lift links paired with appropriately sized beams show that this width of bearing is a reasonable lower bound value that is suitable for design use. For practical design application, the values of k_1 and t_f used to calculate the width of bearing must be based on an appropriate "typical" beam size selected by the link designer.

The limit state bearing stress between the lift link and the beam may be taken as the same value as that used for pin bearing, again based on the allowable bearing stresses given in AISC (1989) and AISC (2010b). In this case, the bearing load should be based on the lesser of the yield stress of the lift link or the yield stress of the beam and is calculated using Eq. 6.52.

$$P_b = 1.50 F_y W_b t \tag{6.52}$$

Application of a design procedure using these limit state strength formulations is best described by means of an example problem. Example 6-4 proceeds through the design of a flat plate lift link of a style commonly used in the industry. The material specification and required dimensions for the strength calculations are shown in Fig. 6.28.

EXAMPLE 6-4

Lift link dimensions and material as shown in Fig. 6.28
Rated load = 100 tons (90.7 tonnes)
Header beam shape - W14x233,
    for which t_f = 1.72" (43.7 mm) and k_1 = 1.75" (44.5 mm)
Header beam F_y = 36 ksi (248 MPa)

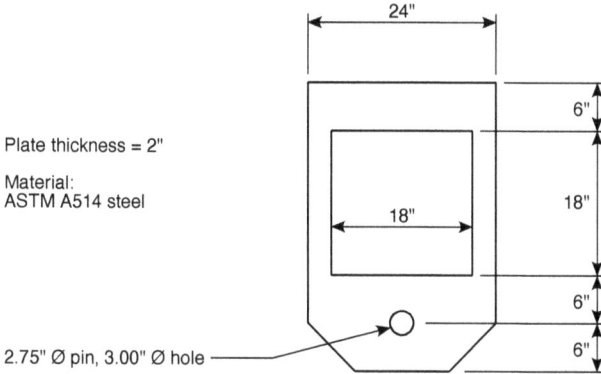

Figure 6.28 Lift Link for Example 6-4

The upper segment of the link (Fig. 6.29) is analyzed to determine its limit state load.

The plastic limit of the upper segment is reached when hinges form at the ends $(T + M)$ and at either the end of the width of bearing $(V + M)$ or on the centerline (M only). A simple method of solving this problem calls for assuming a limit state load, checking the three critical sections, and then adjusting the load until the limit is reached. This can be done manually or through programming of a spreadsheet or mathematics software package. For this example, assume a limit state load of 495.87 kips.

M_p = 100 x (2.00 x 3.00^2 / 4) = 450.00 kip-inches
T = 495.87 / 2 = 247.935 kips
T_y = 100 ksi x (2.00" x 3.00") = 600.00 kips
Eq. 6.47 can be solved with $V = 0$ to determine the moment M at which this section reaches full plasticity.

$$1 = \frac{M}{M_p} + \left(\frac{T}{T_y}\right)^2 = \frac{M}{450.00} + \left(\frac{247.635}{600.00}\right)^2$$

M = 373.16 kip-inches
Width of bearing W_b = 2 k_1 + 2 t_f = 2 x 1.75 + 2 x 1.72 = 6.94"

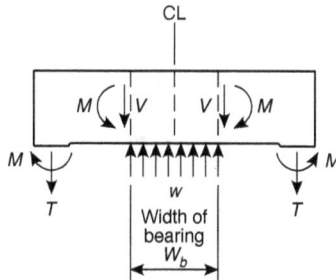

Figure 6.29 Lift Link Upper Segment for Example 6-4

CL end section to edge of width of bearing = 3.00" / 2 + 18.00" / 2 - 6.94" / 2 = 7.03"
Thus, at this section:
M = 247.935 x 7.03 - 373.16 = 1,369.82 kip-inches
M_p = 100 x (2.00 x 6.00² / 4) = 1,800.00 kip-inches
V = 247.935 kips
V_y = 100 / $\sqrt{3}$ x (2.00 x 6.00) = 692.82 kips

Eq. 6.47 is solved using these values and T = 0.

$$\frac{M}{M_p} + \left(\frac{V}{V_y}\right)^4 = \frac{1,369.82}{1,800.00} + \left(\frac{247.935}{692.82}\right)^4 = 0.78 < 1.00$$

Since the interaction ratio is less than 1.00, this section is not fully plastic.

Last, the moment on the centerline is calculated and compared to M_p.
w = 495.87 / 6.94 = 71.451 kips per inch
CL end section to link CL = 3.00" / 2 + 18.00" / 2 = 10.50"
M = 247.935 x 10.50 - 71.451 x (6.94 / 2)² / 2 - 373.16 = 1,799.99 kip-inches

The centerline moment is found to be equal to the plastic moment. Thus, the upper segment becomes fully plastic at a load of 495.87 kips.

This process is repeated for the lower segment (Fig. 6.30). Here, the limit state load is the load that results in a fully plastic section at each end and yielding on the two planes below the pin hole. A limit state load of 876.59 kips (438.295 kips per side) is assumed and Eq. 6.47 is solved as shown for the upper segment to find an end moment M of 209.87 kip-inches in the fully plastic condition.

The centerline moment is:
M = 438.295 x 10.50 - 209.87 = 4,392.23 kip-inches

The net area above and below the pin hole is 2.00 x (6.00 - 3.00 / 2) = 9.00 in.² and the center-to-center vertical spacing of these two areas is (6.00 - 3.00 / 2) + 3.00 = 7.50". Thus,

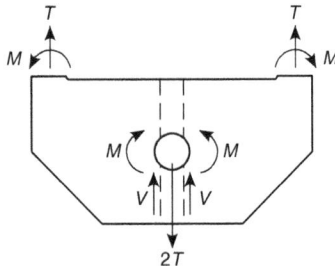

Figure 6.30 Lift Link Lower Segment for Example 6-4

the tension in the area below the pin hole due to the moment is:

$T = 4,392.23 / 7.50 = 585.63$ kips

$T_y = 100 \times 9.00 = 900.00$ kips

The shear area below the pin hole is calculated using Eqs. 6.48 and 6.49.

$$\phi = 55\frac{D_p}{D_h} = \frac{2.75}{3.00} = 50.42°$$

$$Z = a + \frac{D_p}{2}(1 - \cos\phi) = 4.50 + \frac{2.75}{2}(1 - \cos 50.42) = 5.00"$$

$A_v = 2.00 \times 5.00 = 10.00$ in.^{2}

$V_y = 100 / \sqrt{3} \times 10.00 = 577.35$ kips

$V = 438.295$ kips

The area below the pin hole is now evaluated for the combination of shear and tension using Eq. 6.47 with $M = 0$

$$\left(\frac{T}{T_y}\right)^2 + \frac{(V/V_y)^4}{1-(T/T_y)^2} = \left(\frac{585.63}{900.00}\right)^2 + \frac{(438.295/577.35)^4}{1-(585.63/900.00)^2} = 1.00$$

Fully plastic sections are found at end ends and center, thus showing that the limit state load of the lower segment is 876.59 kips.

The lesser of these two values, 495.87 kips, is the strength limit state load of the lift link.

The pin bearing stress limit state is calculated with Eq. 6.51

$$P_b = 1.50F_yD_pt = 1.50 \times 100 \times 2.75 \times 2.00 = 825.00 \text{ kips}$$

Last, the bearing capacity between the lift link and the supporting beam is calculated with Eq. 6.52

$$P_b = 1.50F_yW_bt = 1.50 \times \min(36,100) \times 6.94 \times 2.00 = 749.52 \text{ kips}$$

Both of these bearing limit states are greater than the strength limit state load, so the strength limit state of 495.87 kips governs the design.

The lift link design illustrated in Example 6-4 is based on support on a single-web beam, as detailed in Fig. 6.29. The analysis of the upper segment of a lift link supported on a box section with two webs is based on the configuration shown in Fig. 6.31. Note that the width of bearing over each web may be limited by the width of the flange beyond the centerline of the web to the outside. The locations of the critical planes along the length of the segment will vary, based on the positions of the webs relative to the length of the segment. Last, modeling of the support provided by another type of surface, such as a side shift device of the

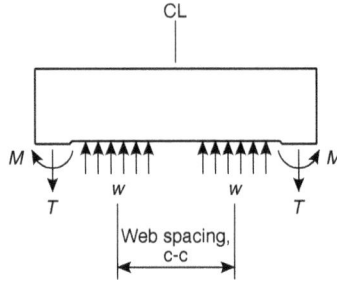

Figure 6.31 Upper Segment Model for Lift Link Support on a Box Beam

type seen in Fig. 2.10, must be evaluated based on the configuration and stiffness of that surface.

The solution presented in Example 6-4 presumes that each segment of the lift link is capable of achieving a full plastic moment without buckling. The stability of each segment (upper, lower, and side) is checked as follows. This method is based on the provisions in AISC (2010b) for the design of a rectangular bar bent about its major axis.

The full plastic moment can be reached for a segment with $\dfrac{Ld}{t^2} \le \dfrac{0.08E}{F_y}$.

For each segment for which $\dfrac{0.08E}{F_y} < \dfrac{Ld}{t^2} \le \dfrac{1.9E}{F_y}$, the plastic moment calculated with Eq. 6.46 must be reduced by the correction factor C_p given by Eq. 6.53, where C_b is given by Eq. 6.54.

$$C_p = \frac{C_b}{1.5}\left[1.52 - 0.274\left(\frac{Ld}{t^2}\right)\frac{F_y}{E}\right] \le 1.00 \qquad (6.53)$$

$$C_b = \frac{12.5\,M_{max}}{2.5\,M_{max} + 3\,M_A + 4\,M_B + 3\,M_C} \qquad (6.54)$$

where
L = segment length measured along centerlines;
d = segment depth;
M_{max} = absolute value of the maximum moment in the segment;
M_A = absolute value of the moment at the quarter point of the segment length;
M_B = absolute value of the moment at the center of the segment length;
M_C = absolute value of the moment at the three-quarter point of the segment length; and,
all other terms are as previously defined.

This plastic analysis method is not recommended if the slenderness ratio of any segment in the lift link is greater than 1.9 E / F_y. Given the practical proportions of plate lift links, such a condition will rarely, if ever, occur.

The lift link design method derived here is based on the ability of the material from which the lift link is fabricated to allow the critical sections to reach full plasticity in the indicated combinations of tension, bending, and shear. Thus, this method may only be used with relatively ductile materials. Most grades of structural steel plate have the necessary level of ductility to perform as needed, but the designer must be aware of this need when specifying material. As previously noted, gantry system lift links are used relatively infrequently, such that the number of load cycles to which any link will be subjected over its life will be small. Thus, fatigue is generally not a concern in the design of lift links and the hot spot stresses that occur in use (and that are seen in finite element analyses of lift links) are not a practical concern. At high loads, some local yielding will occur, but this will not diminish the limit state capacity of the link.

This lift link design method is based on the simple flat plate lift link. While this style of lift link is very common, this isn't the only type of link that the gantry system user will encounter in practice. Some lift links are manufactured as built-up fabrications. The built-up plates may be circular cheek plates around the pin hole added to provide additional pin bearing or shear area, or they may be more extensive doubler plates added to increase the bending strength of one or more elements of the lift link (Fig. 6.32). Regardless of the specific details of the added plates, the fundamental design concept presented here is still applicable. It just becomes a matter of locating the critical sections of the built-up lift link and calculating the limit state capacity accordingly.

The last issue to be considered is the design factor. We saw in Chapter 4 that the upper bound vertical load that may occur at any lift point during the performance of a lift may be twice the design load as a result of cross cornering. Thus, a design factor somewhat greater than 2.00 is appropriate. It is also noted that lift links may be subjected to transverse loads as a result of the various horizontal loading effects discussed in Chapter 4. As shown in Table 4.1, cross cornering most likely will occur only when raising or lowering a load. Thus, the large potential increase

Figure 6.32 Built-up Plate Lift Link *(J&R Engineering Co., Inc.)*

in the vertical load applied to the lift link generally will not occur in combination with transverse loading. Noting the comparatively small horizontal loads, we can see that designing a lift link for a simple straight pull and providing a design factor to address the overloading potential from cross cornering is adequate for normal link design.

As with load combinations and factors as discussed in Chapter 4, establishing design factors for specialty components such as lift links is the domain of standards-writing bodies. We can conclude from this discussion, however, that a suitable design factor for lift links is something greater than 2.00, but not necessarily a lot greater. A value on the order of 2.25 with respect to the limit state load calculated by the method proposed here appears to be reasonable. In the absence of guidance from a standard, the lift link designer or manufacturer must decide on a suitable design factor.

6.7   PROPEL DESIGN

Gantry leg propel mechanisms, whether they are internal drives, external drive wheels, or external cylinders, are almost always provided by the gantry manufacturer as a part of the system. Design of these components is rarely within the scope of responsibility of the gantry system user. The most likely situation where a gantry user may have the need to design a propel device is if the system requires a unique configuration of external propel cylinders for a particular project. In any case, though, it is useful for the gantry system user to understand the demands placed on the propel system and how these systems function.

The design of the propel system is quite simple. The engineer calculates the motive force required to move the gantry system along the track and then selects mechanical and/or hydraulic components capable of supplying that force. The required motive force is equal to the rolling resistance of the gantry leg plus the grade resistance created by the permissible out-of-level of the track in the longitudinal direction plus some additional force required to accelerate the mass of the system.

We found in Section 4.1.3 that the rolling resistance of a gantry leg has been determined experimentally to be about 1.3%. The longitudinal out-of-level tolerance for gantry track is limited to 0.125 inch in 10 feet (3.1 mm in 3 meters) in SC&RA (2004). This is a slope of 0.1%. A reasonable upper bound force required to accelerate a loaded gantry system from rest to its travel speed was shown in Section 4.1.4 to be about 1.0%. Thus, the total resistance to motion of one gantry leg is (theoretically) equal to the sum of these three values, or 1.3% + 0.1% + 1.0% = 2.4%. Standard practice calls for designing the propel system for some greater resistance value to account for inefficiencies and unusual conditions that may occur in the field during the operation of the gantry system.

A gantry system with internal drives most commonly has a complete drive assembly in each leg. Based on this arrangement, the design of each drive need only consider the dead weight and supported load of one leg. Gantry systems with external drive wheels or external propel cylinders (Fig. 6.33) are often configured such that one propel device is required to move two gantry legs through the use of a connecting member. For example, consider a gantry leg that weighs 6,000 pounds and has a maximum rated load of 100 tons. If we use a propel design resistance of 3% of the total load, the design motive force required to move one leg while supporting the maximum load is 0.03 x (100 x 2,000 + 6,000) = 6,180 pounds. For an arrangement in which one propel device must move two legs, the propel design load becomes 2 x 6,180 = 12,360 pounds.

Internal and external drives generally use hydrostatic motors and either a roller chain drive or a gear drive to connect the motor's output shaft to the wheels. The drive torque required at the wheels is simply the design motive force multiplied by the radius of the wheel. To continue with the above example, if the design motive force is 6,180 pounds, the radius of the gantry leg's wheels is 5 inches and only one set of wheels on one axle is driven, then the required torque to be applied to that drive axle is 5 inches x 6,180 pounds = 30,900 inch-pounds.

Also to be determined is the drive speed. The maximum travel speed of a loaded gantry leg should be very slow, perhaps no greater than about 10 feet per minute (3.05 meters per minute). A wheel with a radius of 5 inches has a circumference of π x (2 x 5 inches) = 31.42 inches (798 mm), so the maximum rotational speed of the drive wheels should be 10 feet per minute x 12 inches per foot / 31.42 inches per revolution = 3.82 revolutions per minute.

With these requirements of torque and speed established, a suitable hydrostatic motor and drive train can be selected. The demands of a gantry leg propel system are such that off the shelf components normally suffice.

Selection of a propel cylinder is similarly straightforward. The only additional value that must be considered is the angle of the cylinder to horizontal. The greatest angle occurs when the cylinder is fully retracted. Again using the above example,

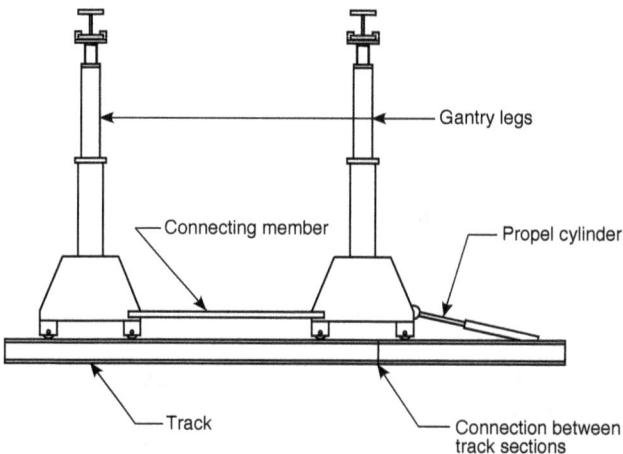

Figure 6.33 Gantry External Cylinder Propel System

the propel cylinder seen in Fig. 6.33 is at an angle of 12 degrees as drawn. Given the required motive force of 12,360 pounds (to move two gantry legs), the required capacity of the cylinder is 12,360 / cos 12° = 12,636 pounds. In order to move the legs in either direction, the cylinder must have this capacity in both extend and retract modes. The speed of cylinder extension and retraction is controlled by means of limiting the fluid flow to the cylinder. The calculations used to equate cylinder dimensions, fluid flow rate, and cylinder extension or retraction speed are discussed in Section 3.3.5.

As with drive wheel propel systems, propel cylinders are also typically off the shelf components. In the event that a specially designed propel cylinder is required, the methods discussed in Chapter 3 provide the tools to establish the maximum fluid pressure and flow rate, equations for calculating the barrel wall stresses due to the internal pressure can be found in Young, et al (2012), and the industry-standard method for calculating the buckling strength of a single stage cylinder supported at its ends is defined by NFPA (2010). A design factor of 2.00 with respect to the strength limit states of barrel wall yield stress and buckling strength is appropriate for establishing the rated load of the cylinder for propel service.

The connecting member, including its end connections to the gantry legs, must be designed to transmit the maximum motive force required to move one leg in either direction. Thus, these members must be designed to carry this force acting in either tension or compression.

The design of the brackets, bolts, pins, and other hardware used to mount the propel components to the gantry leg typically follow conventional structural and mechanical engineering methods. Consideration must always be given to the rigidity of the mounts, as well as to strength. That is, the mounts for the propel components must be stiff enough to maintain alignment of the components during operation. This is particularly important for propel systems that use gear drives or roller chain drives, where misalignment due to excessive flexing of the mounts can result in accelerated wear of the parts and, possibly, premature failure. The designer must also address safety issues, such as providing covers and guards as may be required to protect personnel from injury.

6.8 REFERENCES

Aluminum Association (AA) (2010), *Specification for Aluminum Structures*, Arlington, VA.

American Institute of Steel Construction (AISC) (1989), *Specification for Structural Steel Buildings – Allowable Stress Design and Plastic Design*, Chicago, IL.

American Institute of Steel Construction (AISC) (2010a), AISC 303-10 *Code of Standard Practice for Steel Buildings and Bridges*, Chicago, IL.

American Institute of Steel Construction (AISC) (2010b), AISC 360-10 *Specification for Structural Steel Buildings*, Chicago, IL.

American Society of Mechanical Engineers (ASME) (2012), ASME BTH-1-2011 *Design of Below-the-Hook Lifting Devices*, New York, NY.

APA – The Engineered Wood Association (formerly the American Plywood Association) (APA) (1998), *Plywood Design Specification*, Tacoma, WA.

American Wood Council (AWC) (2012), *National Design Specification (NDS) for Wood Construction*, Leesburg, VA.

Chen, W.F., and Newlin, D.E. (1973), "Column Web Strength in Beam-to-Column Connections." *Journal of the Structural Division*, Vol. 99, No. ST9, American Society of Civil Engineers, Reston, VA.

Crane Manufacturers Association of America, Inc. (CMAA) (2010a), Specification #70 – *Specifications for Top Running Bridge & Gantry Type Multiple Girder Electric Overhead Traveling Cranes*, Charlotte, NC.

Crane Manufacturers Association of America, Inc. (CMAA) (2010b), Specification #74 – *Specifications for Top Running & Under Running Single Girder Electric Traveling Cranes Utilizing Under Running Trolley Hoist*, Charlotte, NC.

Duerr, D. (2006), "Pinned Connection Strength and Behavior," *Journal of Structural Engineering*, Vol. 132, No. 2, American Society of Civil Engineers, Reston, VA.

Duerr, D. (2010), "Effective Bearing Length of Crane Mats," presented at the Crane & Rigging Conference, Houston, TX, Maximum Capacity Media, LLC, Fort Dodge, IA; available for download at http://www.2DM.us/CraneMats.htm.

Elgaaly, M., and Salker, R.K. (1991), "Web Crippling Under Local Compressive Edge Loading." *Proceedings of AISC National Steel Construction Conference*, American Institute of Steel Construction, Chicago, IL.

Federation Europeenne de la Manutention (FEM) (1983), FEM 9.341 *Rules for the Design of Series Lifting Equipment; Local Girder Stresses*, Brussels, Belgium.

Fisher, J.W., and Pense, A.W. (1987), "Experience with Use of Heavy W Shapes in Tension," *Engineering Journal*, Vol. 24, No. 2, American Institute of Steel Construction, Chicago, IL.

Graham, J.D., Sherbourne, A.N., Khabbaz, R.N., and Jensen, C.D. (1959), *Welded Interior Beam-to-Column Connections*, American Institute of Steel Construction, Chicago, IL.

Hannover, H.O., and Reichwald, R. (1982), "Lokale Biegebeanspruchung von Träger-Unterflanschen" ("Local Flexural Stressing of Girder Lower Flanges"), *F+H - Fördern und Heben (Conveying and Lifting)* 32, Nr. 6 [Teil 1 (Part 1)] und Nr. 8 [Teil 2 (Part 2)], Mainz, Germany.

National Fluid Power Association, Inc. (NFPA) (2010), NFPA/T3.6.37 R1-2010 *Hydraulic Fluid Power — Cylinders — Method for Determining the Buckling Load*, Milwaukee, WI.

Neal, B.G. (1961), "The Effect of Shear and Normal Forces on the Fully Plastic Moment of a Beam of Rectangular Cross Section," *Journal of Applied Mechanics*, Vol. 28, No. 2, American Society of Mechanical Engineers, New York, NY.

Roberts, T.M. (1981). "Slender Plate Girders Subjected to Edge Loading." *Proceedings*, Part 2, Vol. 71, Institution of Civil Engineers, London, U.K.

Summers, P.B., and Yura, J.A. (1982), *The Behavior of Beams Subjected to Concentrated Loads*, PMFSEL Rep. No. 82-5, University of Texas, Austin, TX.

Specialized Carriers & Rigging Association (SC&RA) (2004), *Recommended Practices for Telescopic Hydraulic Gantry Systems*, Centreville, VA.

Timoshenko, S.P., and Gere, J.M. (1961), *Theory of Elastic Stability*, 2nd edition, McGraw-Hill Companies, Inc., New York, NY.

Wilson, W.M. (1929), "Rolling Tests of Plates," Bulletin No. 191, University of Illinois Experiment Station, Urbana, IL.

Wilson, W.M. (1934), "The Bearing Value of Rollers," Bulletin No. 263, University of Illinois Experiment Station, Urbana, IL.

Young, W.C., Budynas, R.G., and Sadegh, A.M. (2012), *Roark's Formulas for Stress and Strain*, 8th ed., McGraw-Hill Companies, Inc., New York, NY.

7 Lift Planning and Operations

The planning and performance of a lift using a telescopic hydraulic gantry system requires an understanding of the basic rigging practices that are common to all industrial and construction lifting operations as well as knowledge that is specific to the use of gantries. The necessary gantry-specific knowledge includes understanding how the gantry system functions, how the hydraulic components operate, the nature of the loads and forces that may act on the gantry during a lift, how to design (or select) header beams and track, how to plan a lift, how to set up and dismantle the gantry system, and, finally, how to perform the lift.

As with any lifting and rigging work, the greatest safety and efficiency are obtained through a combination of competent planning and skillful execution of the lift. The technical issues of gantry system performance and lift engineering are covered in the previous chapters. This chapter looks at the more practical aspects of the planning and execution of a lift made with a hydraulic gantry system. That is, this chapter is where the rubber meets the road.

7.1 EQUIPMENT SELECTION AND EVALUATION

The planning required for making a lift with hydraulic gantries uses a logical progression of equipment selection and design that first addresses the layout and then verifies the acceptability of each piece of equipment. When working indoors, the equipment arrangement can be the most important factor in the job. Tight clearances to the building structure and existing equipment or the need to set the gantry tracks only over floor framing hard points can dictate the basic layout of the system. Therefore, a preliminary equipment arrangement drawing should be prepared as the first step in the planning to identify these limits.

Designing the layout of a gantry system is often an iterative process. An initial layout is created, rigging to the lifting attachments on the load is selected, and the function of the equipment throughout the lift is checked. Adjustments are then made to address clearance issues, equipment capacities, and other performance requirements until the system can make the lift safely and efficiently. When the lift is very simple, such as a basic lift with no horizontal movement, the first layout often does it. A more complex operation, such as a lift that involves travel and side shifting in a confined work space, may require multiple revisions to arrive at the optimum solution.

The various considerations that go into establishing the layout of a gantry system are discussed in this section. Because of the potentially iterative nature of this part of the lift planning, the order of the following subsections is not necessarily the order in which the work is actually performed. The order here is simply that which lends itself to clear presentation.

7.1.1   The Initial Big Picture

A practical first step typically is to make a rough layout of the main components of the gantry system. We will use the layout shown in Fig. 7.1 to illustrate this idea and to show where things can go awry when this first step gets the planning off on the wrong foot.

The task for the gantry system shown in Fig. 7.1 was to receive a press frame in the indicated delivery area inside a plant building, lift the frame from four points and side shift it to a position over the pit, upend the frame so it was hanging from the header beam that spans between Gantry Legs 3 and 4, and then mount the frame onto the machine being assembled in the pit. This is the layout used on an actual lift performed in the mid-1990s.

We can see that the wall at the top of the diagram restricts placement of the track for Gantry Legs 1 and 4. The placement of the track for Legs 2 and 3 was dictated, in part, by the need to provide clearance for a truck to enter the delivery area. The header beams overhang Legs 2 and 3 simply because beams of the illustrated length are what the contractor had available for the job. Last, the control and power unit was placed on an area of the floor in a location near the gantry system, but clear of access to the delivery area.

One important consideration that is not explicitly illustrated in the figure is the routing of the hydraulic lines from the control and power unit to the four gantry

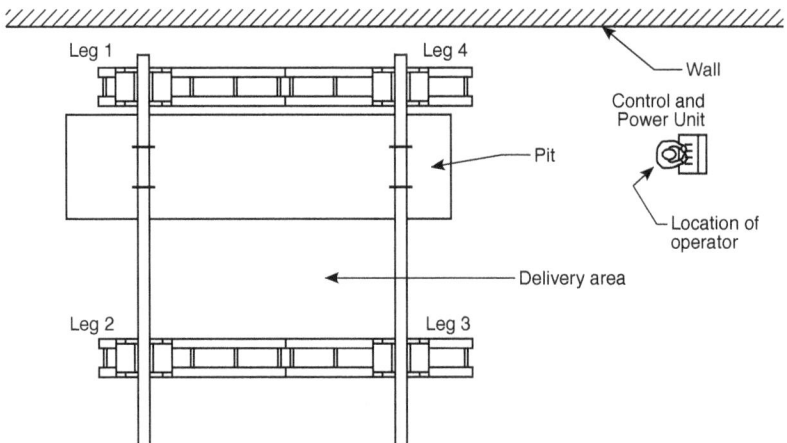

Figure 7.1 Gantry Setup for Lift of a Press Frame

legs. On this job, the truck carrying the press frame has to move into position under the header beams, approaching from the right side of the diagram. This means that the lines from the control and power unit to Legs 2 and 3 either must be protected from the truck's tires (if the lines are laid on the floor) or they must be removed to allow the truck to pass and then reinstalled.

Also related to the control and power unit is exhaust. The power units of many gantry systems are driven by gasoline or diesel engines. When a system of this type is used indoors, as is the case with this example project, provisions must be made to vent the power unit's exhaust to the outside of the building.

The last check that must be made for this initial layout is that of height. The building in which this lift was made has an overhead bridge crane, as well as the usual lights, HVAC ducts, and other overhead building equipment. Clearances between the header beams and side shifting lift links to the building elements must be checked at the maximum lift height and at the extremes of travel of the side shift devices. In this respect, this lift is simpler than many in that the gantry system was not required to travel along the tracks. When travel is required, clearances in three dimensions must be examined to assure that no part of the gantry system or the lifted load will snag on anything throughout the lift.

Now that we have the basic gantry system layout in place, we can examine the steps required to refine the layout into final form.

7.1.2   Rigging, Lift Links, and Header Beams

With the preliminary equipment layout made, the rigging must be sized, lift links selected, and the header beams designed. As discussed in Section 6.1, the selection of slings and rigging hardware for gantry lifting is no different than for crane lifting. Slings, shackles, straps, and the like are used to attach the load to the header beam or cross beam lift links. All of the normal precautions must be observed when selecting the rigging:

- Rig the slings only to acceptable lifting attachments on the load.
- If the load is to be upended or downended, verify that there is clearance between the rigging and the load at all points in the operation.
- Check the clearances above and below the load at all points during the operation.

Under most circumstances, the rigging between the header beams and the load should be as short as is practical. The reason for this is simple. The longer the rigging, the higher the gantry legs will have to be extended to lift the load and the higher the gantry legs are extended, the lower will be the stability of the system. There are downsides to this rule, however. For example, some lifts may require the ability to drift the suspended load sideways to achieve, for example, alignment of bolt holes over anchor bolts when setting the load. Although this needed drift should be small, perhaps less than an inch or so, drifting a load applies a side load to the gantry system.

As an example, consider the two arrangements shown in Fig. 7.2 (the proportions are exaggerated to illustrate the conditions). In both cases, the load must be drifted one inch (25 mm) for setting. The rigging length, measured from the lift link bearing on the header beam to the load's lifting attachment, in Fig. 7.2a is 184 inches (4,674 mm) and the rigging length in Fig. 7.2b is 90 inches (2,286 mm). The force required to drift the load is approximately equal to the weight of the load multiplied by the required drift divided by the rigging length. Thus, the force required for the arrangement shown in Fig. 7.2a is about *Weight* x 1.00 / 184 = 0.0054 = 0.54% of the weight of the load and the force required for the arrangement in Fig. 7.2b is about 1.11% of the weight of the load. We can immediately see that even shorter rigging will result in the need to apply even greater loads to drift a suspended load.

The calculation tools developed in Chapters 4 and 5 provide the lift planner the means to compute the actual side load needed to drift a load in a particular arrangement and to calculate the stability of the gantry system for a specified leg extension. In most cases, the balance of rigging length verses leg extension will be made based on practical considerations of the setup. We can also look to the discussion about lateral loads that result from out-of-plumb rigging as discussed in Section 4.1.3. If the expected misalignment of the rigging will exceed about 1% if the rigging length, then an analysis of the trade-offs between side loading and stability is called for.

Lift links are selected for a particular lift just like the slings and shackles. The applied load is calculated and lift links of a suitable rated load are selected. The lift links must also be sized to fit on the header or cross beams to be used, so we see that lift link selection is also a function of beam sizing. The opening through which the header or cross beam passes should be no more than a few inches (75 mm or so) wider than the beam in order to keep the links close to centered on the beam. If this is not possible (or practical), shims can be used to keep the lift links centered. If the load is to be side shifted, then appropriate side shift devices must be specified.

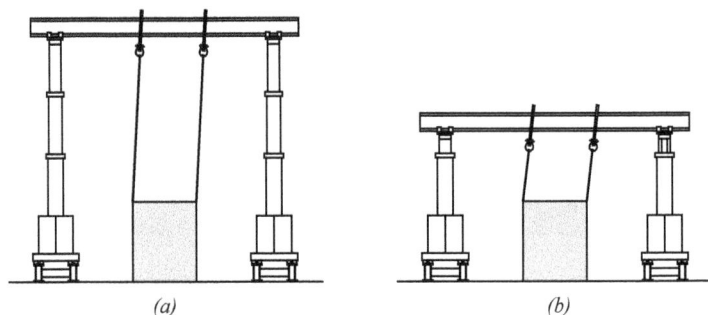

(a)   *(b)*

Figure 7.2 Drifting a Load

Lift links are most commonly purchased from gantry system manufacturers. As such, the lift links are then a vendor-supplied product and have a specified rated load, no different than a sling or shackle. In some cases, however, lifting contractors find it beneficial to design and build their own lift links. One design method applicable to lift links is detailed in Section 6.6.

The design or selection of the header beams (and cross beams, if used) is a relatively simple exercise in structural engineering, the details of which are discussed at length in Section 6.2. Of particular importance in the design of header beams is the selection of impact factors and deflection limits. The SC&RA *Recommended Practices for Telescopic Hydraulic Gantry Systems* (SC&RA 2004) suggests minimum values based on the nature of the lift. Deflection limits are less well defined, with SC&RA (2004) simply recommending that the lift planner follow the advice of the manufacturer of the gantry system being used.

Load charts are available or can be generated that can assist the lift planner in selecting header beams and cross beams for situations where the lifted load is supported by two points per beam. Fig. 7.3 is an excerpt of a simple load chart developed by the author for a wide-flange beam that can be used for either header beams or cross beams. This load chart applies for the simplest case where the two loads are equal and the lift link placement is symmetrical.

Fig. 7.4 is an excerpt of a more elaborate load chart of a type used by one of the gantry manufacturers that shows rated loads based on the limits of both the header beam and the gantry legs. Again, this chart is based on two equal loads, but the load positions need not be symmetrical. "A" and "B" load positions allow locating the two loads independently. The heavy line on this load chart delineates between ratings that are limited by the strength of the header beam and ratings that are limited by the capacity of the gantry legs. (It must be noted that these charts are shown here for illustration purposes only and should not be used by the reader for lift planning purposes.) As noted in Section 6.5, use of load charts like these demands that the user understand the basis for the load ratings shown (i.e., impact factors used, design factors, etc.).

Beam Details...

Shape Designation	W14x211
Depth	15.700 inches
Web Thickness	0.980 inch
Flange Width	15.800 inches
Flange Thickness	1.560 inches
Yield Stress	50 ksi (ASTM A992)

Design Criteria...

Vertical Impact	5.00%
Lateral Impact	10.00%
Allowable Deflection	L / 480

Beam Load Chart...

Span -->	45 ft	40 ft	35 ft	30 ft	25 ft	20 ft	15 ft	10 ft
Load Position								
2.5 ft	158.00	178.50	204.00	239.00	288.00	362.00	492.00	578.00
5.0 ft	80.00	90.50	104.00	123.00	150.00	193.00	266.00	289.00
7.5 ft	54.50	62.00	72.00	86.00	107.00	145.00	177.00	
10.0 ft	42.00	48.00	56.50	69.50	90.00	132.00		
12.5 ft	35.00	40.50	48.50	61.50	85.00			
15.0 ft	30.50	36.00	44.50	59.00				
17.5 ft	28.00	34.00	43.50					
20.0 ft	26.50	33.00						
22.5 ft	26.25							

Figure 7.3 Example Header Beam Load Chart

LIFT BEAM LOAD CHART WITH 300-2-24 GANTRY

DIST.	LOAD PER LEG											

ONE (1) LIFT BEAM
TWO (2) LIFT LEGS
MAXIMUM
LOAD PER LEG

DIST. A IN FT.	37.5 TONS	DISTANCE B IN FEET								
		3.0	4.0	5.0	6.0	7.0	8.0	9.0	10.0	11.0
3.0	LIFTING CAPACITY	72.9		LIMITED BY		LIMITED BY				
	LEG A LOAD	37.5		GANTRY		BEAM				
	LEG B LOAD	37.5		CAPACITY		STRENGTH				
4.0	LIFTING CAPACITY	69.7	68.6							
	LEG A LOAD	34.3	35.4							
	LEG B LOAD	37.5	35.4							
5.0	LIFTING CAPACITY	60.4	57.5	54.9						
	LEG A LOAD	28.5	28.5	28.5						
	LEG B LOAD	34.0	31.1	28.5						
6.0	LIFTING CAPACITY	53.0	50.3	47.9	45.7					
	LEG A LOAD	23.9	23.9	23.9	23.9					
	LEG B LOAD	31.2	28.5	26.1	23.9					

Figure 7.4 Combined Load Chart for Gantry Legs and Header Beam

7.1.3   The Gantry Legs

The next items to be selected are the next items in the load path: the gantry legs. This should be a very simple task, since the gantry legs are a manufactured system with load charts, an operator's manual, and everything else that the user needs to understand how to use the system. Unfortunately, experience shows that it isn't quite so simple.

Just as a 400-ton mobile crane can rarely actually lift 400 tons, a 400-ton gantry system often cannot lift 400 tons. A 400-ton gantry system can only lift 400 tons when the load is shared exactly equally among all of the legs of the system. Consider, for example, the gantry system load chart of Fig. 7.5. This is the load chart for an 800-ton capacity four-leg system. As we can see, the maximum rated load of the system is 800 tons in the first stage. However, what happens when the load is not shared equally among the four legs?

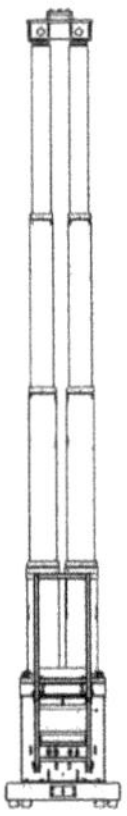

PRESSURE	2,000 PSI (138 BAR)	1,800 124	1,600 110	1,400 97	1,200 83	1,000 69	800 55	600 41	400 28	200 14
27' 5" (8,357 mm) 3rd STAGE	400 TON (362)	361 TON (327)	321 TON (291)	281 TON (254)	241 TON (218)	201 TON (182)	160 TON (145)	120 TON (108)	80 TON (72)	40 TON (36)
21' 7" (6,579 mm) 2nd STAGE	597 TON (541)	537 TON (487)	477 TON (432)	418 TON (379)	358 TON (324)	298 TON (270)	238 TON (215)	179 TON (162)	119 TON (107)	59 TON (53)
15' 9" (4,801 mm) 1st STAGE	800 TON (725)	747 TON (677)	664 TON (602)	581 TON (527)	498 TON (451)	415 TON (376)	332 TON (301)	249 TON (225)	166 TON (150)	83 TON (75)
9' 11" (3,023 mm)									STANDARD	(METRIC)

Figure 7.5 Gantry System Load Chart

Consider the lift arrangement detailed in Fig 7.6. The load to be lifted weighs 60 tons, but it is offset to one side. The distance from the centerline of the left gantry leg to the center of the load is one-third of the header beam span and the distance from the center of the load to the right gantry leg is two-thirds of the span. The sharing of the lifted load between the two gantry legs exhibits this same one-third / two-thirds ratio. That is, the left leg will bear two-thirds of the weight of the lifted load, or 40 tons, and the right leg will carry only one-third of the weight, or 20 tons.

We can see by this example why the 800-ton system often cannot actually lift 800 tons. More importantly, we can see the importance of selecting the gantry system on the basis of the rated load *per leg*, and not the rated load of the complete system. We can also see the utility of load charts such as that shown in Fig. 7.4. This chart takes into account the division of the lifted load between the two legs and the rated load per leg.

When presented with a load chart for the complete gantry system, such as that shown in Fig. 7.5, the lift planner must determine the rated load of one leg. To do this, simply divide the chart rated load by the number of legs (in this case, 4). For example, the rated capacity of one leg operating at a hydraulic pressure of 1,800 psi and extended in the second stage is 537 / 4 = 134.25 tons.

The example of Fig. 7.6 shows the lifted load fixed in one offset position. If the load is to be side shifted along the header beams while suspended, the maximum load to each leg throughout the range of side shifting must be determined to be sure that neither leg will be overloaded.

The load chart of Fig. 7.5 illustrates another important facet of gantry system selection: lift height. The rated load of most gantry legs varies from one stage of extension to the next, with the rated load dropping as each successive stage of the

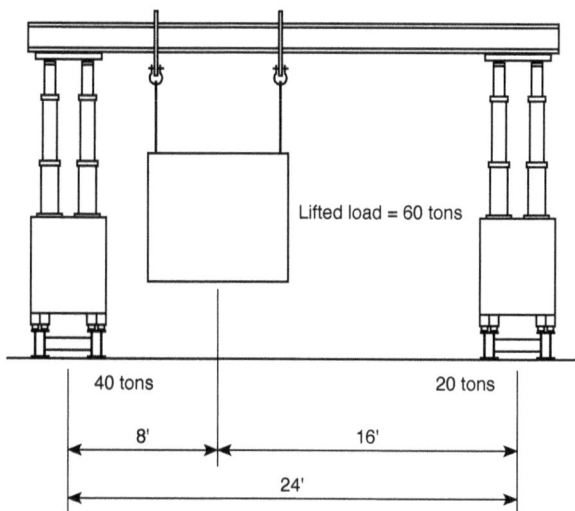

Lifted load = 60 tons

40 tons   20 tons

8'   16'

24'

Figure 7.6 Load Distribution with an Offset Lift

lift boom or lift cylinder is extended. Thus, the lift planner must take a careful look at the heights to which the legs must extend to make the lift.

The check of lift height encompasses an examination of many dimensions and components. For a straight vertical lift, questions to be asked include: Can the gantry legs extend high enough to lift the piece to its required elevation? When the piece is lifted to its highest elevation, is there adequate clearance above the header beams to the roof or any other overhead obstructions? This concern is no different than checking to be sure that a crane doesn't two-block.

If the gantry system is to be traveled along the tracks with the load suspended, do adequate clearances exist from the track up to the tops of the header beams and lift links along the complete path of travel?

When upending or downending a load, clearances between the two sets of gantries and between the load and the header beams must be verified. Rigging lengths and gantry leg extensions must also be checked. An example of an upending operation is illustrated in Fig. 7.7. Points to consider include:

- As the load moves closer to vertical, the lifting and tailing gantries move closer to one another. If the two gantry sets cannot move close enough together, the final upending will pull the slings out of plumb, which will, in turn, impart possibly significant longitudinal forces to the gantry legs.
- As the load moves closer to vertical, the top side of the load approaches the tailing gantry header beam. This potential interference can be relieved by using tailing lugs that extend out from the surface of the load or by using

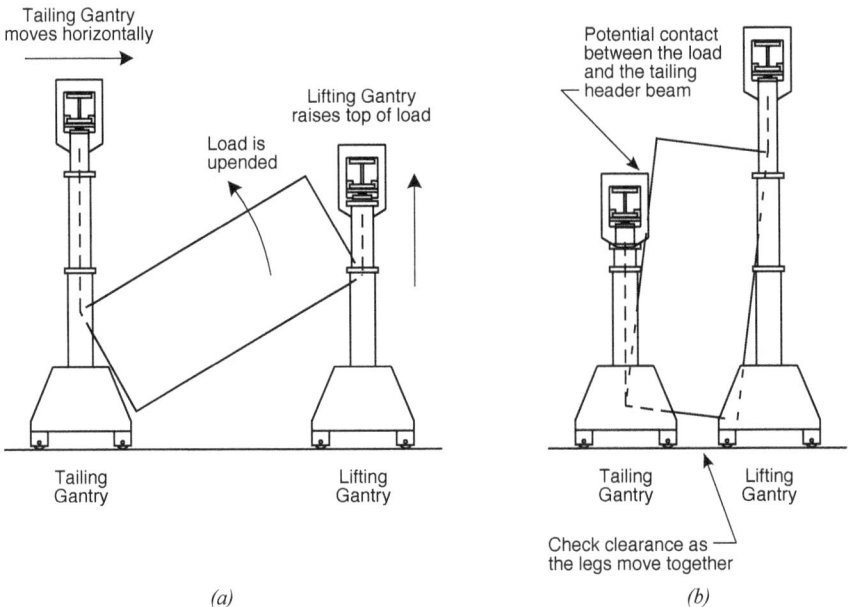

(a)   *(b)*

Figure 7.7 Clearances When Upending a Load

long enough rigging to the tailing lugs that the tailing header beam will remain above the load after the load has been fully upended.

- The lifting gantry must have the stroke needed to keep its header beam above the fully uprighted load.

We can see from this example that planning an upending or downending operation can be a fairly complex undertaking. The analysis of this operation must jointly consider the dimensions of the load, the locations of the lifting attachments, the location of the center of gravity, the lengths of the rigging, the extension heights of the lifting and tailing gantries as the lift proceeds, and the changing rated loads of the gantries at the changing extension heights.

At this point in the planning, the general layout of the equipment has been established, the rigging and lift links selected, the header beams and cross beams designed, and the adequacy of the gantry legs verified.

7.1.4   Evaluation of Site Conditions

The next components of the gantry system that must be selected are the track assemblies and any hardware related to the support of the track. Before that can be done, the support conditions at the site must be evaluated and understood. Seven general track support conditions are discussed here.

When the gantry system is set up outdoors, such as at a rail siding, the system often is supported on soil or light paving, such as asphalt. This type of surface often requires leveling, either by removing material or backfilling. In either case, compaction may be required to improve bearing capacity. Of concern here is not only soil strength, but the tendency of soil to settle under load. Consistent strength and stiffness are needed to safely support the track and to avoid differential settlement as the lift proceeds. Further, underground pipes, catch basins, and other structures may limit the magnitude of the load that can be applied at the surface. The site owner should be consulted for information about subsurface conditions. A geotechnical investigation may be required to accurately evaluate the allowable ground bearing capacity.

A gantry system is often set up on a concrete slab on grade. This can be a ground level floor in a building or a paved area outside. In either case, the strength of the slab must be evaluated to assure that the concentrated loads from the gantry legs will not damage the slab. This evaluation requires knowledge of the slab construction (thickness; grade of concrete; size, grade and placement of reinforcing steel), as well as the condition of the subgrade material on which the slab bears. The presence of voids in the subgrade under the slab can create weak spots and, again, underground pipes and other structures are of concern. Setting any of the gantry legs near the edge of the slab will increase the stresses in the slab and the pressure on the subgrade. The contractor should request that the site owner provide construction drawings to help in the evaluation of a slab on grade. Not surprisingly, experience tells us that the older the facility, the less likely it is that construction drawings will be available.

When working in any type of building structure above ground level, the gantry track beams may be supported on the building's floor. It is unlikely that a conventional floor structure will be able to safely support the high loads from a gantry system. In this situation, structural drawings of the building must be obtained and the assistance of a structural engineer sought to fully check the framing. If the structure cannot safely support the gantry system loads, shoring may be required.

Gantry systems are often used to set large machinery components, such as power generation equipment, on heavy foundations. In these situations, the track may be set directly on the foundation. Due to the typically substantial size and mass of these foundations, solid support is assured and further analysis is generally not required. However, the site survey must include locating all anchor bolts and other embedments that may interfere with track placement.

If the track will bear on the ground surface near a bulkhead or the edge of a pit, the bulkhead or the pit walls must be analyzed for the lateral pressures created by the imposed surface loads. A vertical load applied to the surface of the ground creates lateral pressures in the soil below. The magnitudes of these pressures vary, based on the nature of the soil and the distance of the vertical load to the wall or bulkhead. This behavior is well understood by civil engineers and can be analyzed once some basic site properties are known. The resulting lateral pressures are then used to evaluate the adequacy of the wall or bulkhead.

One often unrecognized area of risk occurs when the track transitions from one type of support to another. This occurs, for example, when the gantry system must travel along track beams that run along bare ground and then pass onto the edges of a large foundation. The concern here is that the soil may settle, but the foundation will not, thus allowing the track to move out of level in the longitudinal direction. Compensation for this behavior can be developed through support placement and sizing, provided that the soil behavior under load is reasonably understood.

Support of a gantry system on a barge is a very specialized application that requires detailed engineering analysis. A free floating barge will move, possibly significantly, as the lifting operation proceeds. A hydrostatic analysis of the barge is required to quantify these motions. Whether the barge is grounded to prevent these motions or free floating, the deck and internal structural members must be checked to assure that they can support the gantry system. And, of course, the barge must be suitably anchored to hold it in position.

These seven site conditions that affect the gantry track layout are the most commonly found. Other conditions may be encountered and must be evaluated appropriately to assure a safe setup. The next section on track selection discusses some of these site conditions and adds detail to how to address some specific problems that are familiar to gantry system users. The consistent theme is recognition of how gantry system operation affects and is affected by the track setup.

7.1.5   Track

The last design item we will discuss here is quite possibly the most important: the track. If the foundation on which the gantry legs are mounted is not firm and level, a complete collapse of the system is possible. The track under one gantry leg is usually a pair of beams. Both beams must be the same size and supported in the same manner if the legs are to be safely supported. When the load is picked, the track beams will deflect. If one track beam deflects more than the other, the gantry leg will lean to the low side (that is, the side with the greater track beam deflection). This will shift the top of the gantry leg to that side, thereby increasing the load to the track beam on that side, thus causing that track beam to deflect even more. If the difference between the deflections of the two track beams becomes great enough, the leg will topple.

The contractor must assure that the track beams under a gantry system are of the same size and similarly supported so that they will deflect equally as the leg moves along the track. However, differences in the loading of one track beam compared to the other can result in differential deflections and, unfortunately, there are always variations that can cause unequal loading. For example, if the two track beams under one leg are not level with each other, the leg will tip toward the low side and more heavily load that beam. Thus, the beam on the low side will deflect more than the beam on the high side. If the track beams are designed for very small deflections under the theoretically equal loading, these minor errors will not be meaningful. The difference in deflections will be a small part of a small number, which, in most cases, will be insignificant. The effect of unequal track beam deflections on gantry leg stability is discussed in detail in Chapter 5.

Such a risk can occur when working at the edge of a pit or other floor opening. Consider, for example, the track beam setup illustrated in Fig. 7.8. In the arrangement shown in Fig. 7.8a, the right track beam is laid directly on the floor at the edge of the pit and the left beam spans over the pit. When loaded, the left track beam will deflect and the right track beam will not. This difference in deflection will cause the gantry leg to lean to the left side.

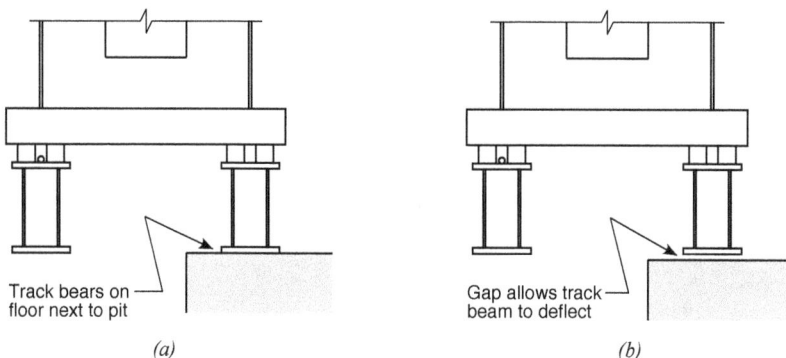

Track bears on floor next to pit

Gap allows track beam to deflect

(a)   (b)

Figure 7.8 Track Beam Support Over a Pit

The solution to this problem is quite simple. Both track beams should be supported on shims, as shown in Fig. 7.8b. The shims that support the right beam must be in line with those under the left beam, such that both beams span the same distance. In this configuration, as the gantry leg moves out onto the span, both track beams will provide the same support stiffness. Assuming that the track is initially level and the leg is straight and plumb, both beams will deflect the same amount and the gantry leg will remain plumb and, thus, stable.

One solution to this problem that does not work well is that of trying to support the left track beam by cribbing up under the beam inside the pit. The problem with this approach is that timber cribbing will crush and deflect more than the solid floor that is supporting the right track beam. Although the difference in vertical deflection will obviously not be as great as that which occurs when the left track beam simply spans the pit, a potentially large and unacceptable deflection will still occur. Supporting the left track beam by more substantial means, such as steel track stands that are tightly shimmed to the underside of the beam, may provide a workable solution. Just keep in mind that the ultimate goal in the placement and support of the track sections is to provide a stiff and uniform support to the gantry legs.

Another potential problem area is illustrated in Fig. 7.9. When the track assembly runs along side a foundation, the possibility exists that one track beam will bear on the relatively immovable foundation and the other beam will bear on the adjacent soil. As with the situation of a track beam spanning a pit, this arrangement results in markedly different stiffnesses between the two beams. There is no easy fix to this problem, other than to avoid it altogether. If clearances do not allow positioning the track assembly such that both beams bear on the concrete foundation, then the assembly can be shifted outward such that both beams bear on the soil.

Last, if the gantry system is set up entirely on soil, additional load spreading may be necessary to reduce the ground bearing pressure to a tolerable level. This

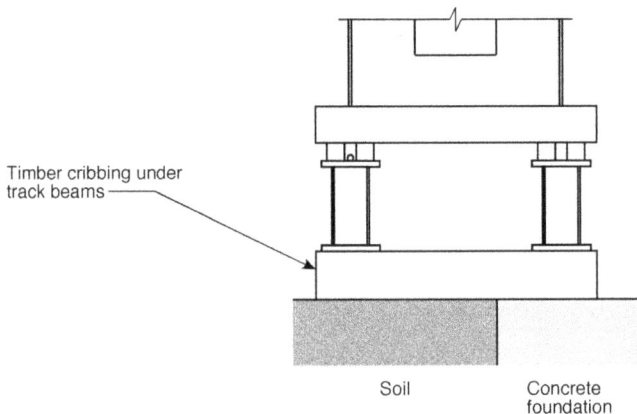

Timber cribbing under track beams

Soil

Concrete foundation

Figure 7.9 Track Beam Support Bearing on Different Surfaces

can be achieved by setting the gantry tracks on standard timber crane mats (Fig. 7.10*a*) or on fabricated steel mats (Fig. 7.10*b*). In either case, the mat strength must be evaluated for load spreading across the mat's width as well as along its length.

As noted before, the track is the foundation upon which the entire gantry system is supported. Exceptional care must be taken when planning the track layout to assure that the final installation will provide the firm and level support needed to allow the gantry system to operate safely.

7.1.6   Location of Accessory Equipment

The layout of the load-carrying components of the gantry system is usually controlled in great part by the demands of the lift and the characteristics of the work site. Consequently, the lift planner will usually focus most on these components. However, the work doesn't end there. There is one more vitally important component that must be positioned correctly for a safe, efficient, and successful lift: the gantry system operator.

Look again at the gantry system layout illustrated in Fig. 7.1. Notice the orientation of the control and power unit. This unit was positioned such that the operator had his back to the lift. (And remember, this layout was used on an actual project.) This shouldn't need to be said, but obviously it does. The gantry system operator must be able to observe the lift, if at all possible.

There are conceivable situations where the four legs of a system may be located around a very large load such that the operator will not be able to view all of the legs from one position. In such a case and when operating with a stationary control unit, the operator must work with one or more spotters to assure that he has eyes on the complete system at all times. When operating with a portable control unit, as is currently becoming more popular, the operator should continually move around the lift to observe the legs, stopping the lift when necessary to assure that

(a)   *(b)*

Figure 7.10 Mats Used for Load Spreading *(Burkhalter Rigging, Inc.; Edwards Moving & Rigging, Inc.)*

all equipment is functioning as required and that the movements of the gantry legs, both vertically and horizontally, are fully coordinated so as to minimize the load effects created by out of sync operations.

The lift that was performed using the setup shown in Fig. 7.1 gives an ideal lesson of what can happen when the guidelines discussed in this section are not followed. Once the press frame had been side shifted to its position over the pit, the upending process began. Gantry Legs 3 and 4 were extended to raise the top of the frame. Due to the significant offset of the load, Gantry Leg 4 was much more heavily loaded than Leg 3. Legs 3 and 4 extended into the third stage, at which point the load on Leg 4 well exceeded its rated load and the leg stalled. Leg 3 continued to extend. Since the operator was facing away from the lift, he didn't see what was happening. The extension of Leg 3 moved the header beam out of level to the point that the side shift devices on this header beam broke free and rolled downhill. The press frame struck Leg 4 and the entire gantry system collapsed. Fortunately, there were no injuries in this accident.

This one event illustrates the need to understand the equipment and to design the gantry system layout to allow a safe execution of the lift.

7.2 DETAILED LIFT PLANNING

As previously noted, performing a lift with a hydraulic gantry system is, in many ways, no different than lifting with a conventional mobile or overhead crane. The lift must be planned from beginning to end, the right equipment must be selected, and appropriately trained people must operate that equipment.

Every lift must be planned. However, this does not mean that every lift requires the preparation of an elaborate lift plan method statement with engineering drawings and calculations. This simply means that the activities required to complete the lift must be thought out and communicated to the people performing the work. For a straight "lift and lower" operation where the lifting equipment is used at a relatively low percentage of rated load, this may be just a rough sketch on a note pad transferred to chalk marks on the floor. A more complex lift may demand the preparation of formal drawings and calculations.

7.2.1 Lift Plan Considerations

Whether the final lift plan is a rough sketch or a formal document with drawings, calculations, and a detailed written method statement, the process of planning a lift using a hydraulic gantry system should typically address, at a minimum, the following items.

1. Define what must be done with the load ("transfer from rail car to trailer," "upend and set on foundation," etc.). This one-sentence summary should

then be expanded to a short narrative that describes the lift from beginning to end ("stage rail car immediately north of grade crossing; stage trailer to the east of the rail car; lift item from rail car, side shift to the east, and set onto trailer").

This seems obvious, but is not always as easy as one might assume. Having to think through the steps that must be performed, even for a basic lift, often results in the realization that important information is missing. Where can the rail car be staged? Will the trailer fit where it has to go? What restrictions exist in the area that will hinder setup of the gantry (e.g., overhead power lines)? This first lift planning step sets the stage for all subsequent work, so spend the time necessary to get it right.

2. Determine the configuration of the load in terms of overall dimensions, weight, center of gravity location, locations and types of lifting attachments (padeyes, trunnions, etc.).

    This information is usually available on vendor drawings and specifications for new items, although added material, such as insulation, can add to the specified weight. Assembling this information for an older item, particularly a much older item, can be challenging. Drawings may not be available, which means a field survey is required to measure the overall geometry of the item and to provide the basis for weight take-off calculations. Some types of equipment, such as process vessels, may be fouled internally with waste material which will increase the lift weight, sometimes significantly, above the value shown on drawings or nameplates. Knowing the weight of the load is fundamentally important for any lifting operation.

3. Locate all obstructions along the path of movement, including doorways, building columns, overhead items, such as roof beams or overhead cranes, and the like.

    As with the load, itself, this information is often available in the form of drawings of the facility. If not, a site survey will be required. If all clearances between the gantry system with the load and the surrounding structures, equipment, etc. are generous, formal action may not be necessary. If clearances are tight, the preparation of plan, elevation and section drawings may be needed to assure that everything will clear throughout all movements of the load and the gantry system. If the gantry system will extend to a height that could interfere with an overhead crane, the overhead crane should be shut down to assure that it cannot be operated inadvertently during the performance of the gantry lift.

4. Locate all below-level obstructions, including pits or floor openings, weak floor framing, underground pipes, and any other items that may control or limit the placement of track.

This information falls into the same category as that discussed above in Item 3 and as in Item 3, drawings of the facility may show much of what is needed here. However, unlike doorways and columns, that is, unlike features that one can readily see, below-level features are often invisible. A concrete floor slab may look substantial, but a void under a section of the slab can reduce its load carrying ability significantly. Likewise, underground pipes or sewers can limit the allowable ground bearing capacity at the surface. In addition to reviewing whatever drawings are available, a site inspection to look for signs of other features not shown on the drawings may be prudent.

5. Make a preliminary selection of equipment and prepare an equipment layout drawing. For a simple lift, this may be nothing more than a hand sketch. For a more complex lift, formal engineering drawings may be required.

    This activity is discussed in Section 7.1.1. This is where we bring together the information gathered in Items 1 through 4 above and begin formulating the actual lift plan.

6. Compute the tensions in the rigging based on the weight of the load being lifted and the locations of the lift points. For a four-point pick, it may be necessary to consider cross cornering.

7. Select slings, shackles, lift links, and other rigging components as required to carry the calculated loads.

    Items 6 and 7 are discussed in Section 7.1.2. In addition to selecting the rigging gear, appropriate documentation may be required in some work environments. Owners of some types of facilities, such as refineries and most any facility in the nuclear industry, require test and inspection certificates for all rigging gear used on their sites. This is a good time to assure that all required documentation is available to be made a part of the final lift plan package.

8. Select appropriate vertical and lateral impact factors and deflection limits (see Chapter 6) and design or select the header beams. The impact values given in the SC&RA *Recommended Practices for Telescopic Hydraulic Gantry Systems* (SC&RA 2004) should be used unless specific job conditions dictate other values.

    When evaluating design loads to a gantry system, wind loading should be considered. As discussed in Section 4.1.6, wind is rarely an issue when lifting with gantries, simply because the systems are not tall and the loads are most commonly compact relative to their weight. However, there are those lifts where wind loading becomes significant, so the lift planner must be prepared with the necessary tools.

The calculation of wind loads can be accomplished by following the provisions of ASCE (2010). The most important value that the lift planner must establish in order to apply these calculations is the maximum wind speed to which the gantry system will be exposed. The appropriate design wind speed for the design of a permanent structure can be found in maps in ASCE (2010). These design wind speeds are reduced by factors given in ASCE (2002) for structures that will be in place for less than five years. The wind load calculation for a lift that will take place over just a few hours is more commonly done by specifying a maximum permissible wind speed, above which the lift cannot proceed.

Once such a wind speed limit is established for the lift, the field supervision is then required to monitor wind conditions to assure that the lift does not proceed if the specified maximum wind speed is exceeded. Depending on conditions, the use of an anemometer to provide an accurate measurement may be necessary. If used, the anemometer must be mounted at the top of the gantry system or in a location of similar height that is unobstructed. Alternately, wind speed can be observed approximately by means of the appearances defined by the Beaufort Scale (Table 7.1). Regardless of the method used, the maximum wind speed for the lift and the method by which the wind speed is monitored must be clearly defined in the lift plan.

9. The header beam reactions are the loads that each gantry leg must support. Verify that the selected gantry model is capable of lifting these loads at the required height.

    As discussed in Section 7.1.3, this evaluation must work with the rated load per leg, not just of the gantry system as a whole, and must use the maximum header beam reactions when the load is side shifted. For the purpose of lift plan presentation, the check of the gantry leg should be shown as a percent utilization (equal to the maximum load supported by the leg divided by the rated load of the leg), as is often done for mobile crane lift planning.

10. Compute the total loads at the bottom of each gantry leg. This is the sum of the maximum header beam reaction plus the weight of the gantry leg.

11. If the track must span an opening, select appropriate vertical and lateral impact factors and deflection limits and check the strength and stiffness of the track beams. Again, the SC&RA (2004) values are suggested unless specific conditions dictate otherwise.

    Refer to Sections 7.1.4 and 7.1.5 for a more detailed discussion of the requirements for track design. The track design must also show the layout of the actual track assemblies. For a simple "lift and lower" operation, one track assembly under each gantry leg may be all that is needed, so

TABLE 7.1 Beaufort Wind Scale

Force	Wind Speed, mph (k/hr)	WMO Classification	Appearance of Wind Effects
0	Less than 1 (2)	Calm	Calm, smoke rises vertically
1	1 - 3 (2 - 6)	Light Air	Smoke drift indicates wind direction, still wind vanes
2	4 - 7 (7 - 11)	Light Breeze	Wind felt on face, leaves rustle, vanes begin to move
3	8 - 12 (12 - 19)	Gentle Breeze	Leaves and small twigs constantly moving, light flags extended
4	13 - 18 (20 - 30)	Moderate Breeze	Dust, leaves, and loose paper lifted, small tree branches move
5	19 - 24 (31 - 39)	Fresh Breeze	Small trees in leaf begin to sway
6	25 - 31 (40 - 50)	Strong Breeze	Larger tree branches moving, whistling in wires
7	32 - 38 (51 - 61)	Near Gale	Whole trees moving, resistance felt walking against wind
8	39 - 46 (62 - 74)	Gale	Whole trees in motion, resistance felt walking against wind
9	47 - 54 (75 - 87)	Strong Gale	Slight structural damage occurs, slate blows off roofs
10	55 - 63 (88 - 102)	Storm	Trees broken or uprooted, considerable structural damage
11	64 - 73 (103 - 117)	Violent Storm	
12	74+ (118+)	Hurricane	

the layout design is not complicated. For a "lift and travel" operation, a number of track assemblies will be laid end to end to provide the runway along which the gantry system will travel. Depending on the nature of the surface on which the track bears, designing this layout may require a certain amount of effort to assure that all of the splices between track beams are soundly supported. A clearly dimensioned track layout drawing may be required to show the field crew exactly what track assemblies are used and where each assembly must be located. Given the importance of the track layout, a little extra effort here is time well spent.

12. Check the strength and soundness of the base on which the track is laid. This must include a check of the possibility of differential deflection if some sections of the track are more rigidly supported than others.

The importance of this activity cannot be overstated. Hydraulic gantry systems are very sensitive to inadequate support conditions. Excessive and/

or differential deflection of the track beams can greatly reduce the stability of the gantry legs, possibly leading to toppling of the system. As noted in the discussion that accompanies Figs. 7.8 and 7.9, the planning of the track layout should be such that both track beams of each track assembly are supported in the same manner. That is, the spans of both beams from one support point to the next should be equal in length and the vertical stiffness of those support points should be very similar, if not identical.

Anything other than the most basic arrangement should be detailed on a drawing. Cribbing timbers, shims, and other track support elements must be clearly dimensioned and materials specified (we don't want the field crew using plywood shims where the track beam reactions call for steel plates). If the track will be supported on soil, the allowable ground bearing pressure must be determined, the strength and stiffness of load spreading equipment (cribbing timbers, mats, etc.) evaluated, and the actual applied pressures calculated.

Shims are often required to level the track beams. The size and material of shims must provide adequate bearing strength between the underside of the track beams and the surface below. Steel plates of various thicknesses are preferable, but hardwood planks are also widely used when a thickness of a couple of inches must be built up. One gantry manufacturer states that plywood is not an acceptable track shim material. However, the author notes that plywood is commonly used as a shim material by contractors in the United States. If the contractor chooses to use plywood for shimming gantry track, it is suggested that the shims be sized such that the compressive stress does not exceed 160 psi (1,100 kPa). If the grade and quality of the plywood is known reliably, additional guidance on allowable compressive stresses can be found in APA (1998).

13. Once all of the selected equipment has been verified as acceptable, go back and perform a final check of the layout to assure that adequate clearances exist between the load and the gantry system and between the gantry system and any surrounding obstructions. Decisions made, for example, when sorting out track support, can alter previously checked clearances at the header beam level.

As with all specialized lifting and rigging, the appropriate planning must go into the job up front to head off any problems. Fundamental to this lift planning effort is a full understanding of the abilities and limitations of gantries. While there are similarities to lifting with mobile or overhead cranes, gantry systems do have their own idiosyncrasies that must be recognized and addressed in the lift plan and in the engineering that supports the lift plan.

The SC&RA *Recommended Practices for Telescopic Hydraulic Gantry Systems* contains extensive discussions of lift planning requirements and performance guidelines. Some contractors find checklists and standard forms useful for the planning of relatively simple lifts. Checklists are also included in the SC&RA guide that can be very helpful to those planning and performing lifts with hydraulic gantries. This booklet should be considered required reading for all who work with hydraulic gantry systems.

7.2.2   What is a Critical Lift?

There are few terms used in lifting and rigging that create as much confusion and debate as the term "critical lift." We all understand that a critical lift, whatever it may be, requires a higher level of planning and care in execution than a routine, or standard, lift (whatever that may be). Defining what constitutes a critical lift, however, remains elusive.

The definition of a critical lift is typically established by company policy. The most common criteria in the definition include the following.

- The weight of the load exceeds a specified threshold.
- The total lifted load exceeds a specified percentage of the lifting equipment's rated load.
- The lifting operation will move the load over operating equipment (this criterion is particularly common in refinery work).
- The item has features or characteristics that make lifting unusually difficult or complex.
- The economic loss due to an accident would exceed a specified dollar amount. This criterion may include consideration of the value of the item being handled, the time required to replace the item, if damaged, and the costs to the project due to a delay.

Additional criteria may be applied, based on the nature of a particular project or the experience of the contractor or facility owner.

A lift classified as critical must be performed in accordance with a detailed lift plan and is typically subject to greater management oversight, during both the planning and the performance of the lift. The purpose of this increased oversight is to reduce risk. Risk is created by uncertainty. That is, we don't know all of the facts completely. The greater the uncertainty, the greater the risk. If we carefully study a situation, we can chip away at those uncertainties by improving our knowledge of the facts. Detailed planning of a lift is a structured method by which we improve our knowledge of the circumstances of the proposed lift, thus (in theory) reducing the risk associated with performing that lift.

So how do we define "critical lift" when discussing a lift to be made with a hydraulic gantry system? Some lift policies with which the author is familiar classify all lifts made with gantry systems to be critical lifts. This answers the question, but not in a very efficient way. Let us examine this question along the lines of lift policies that are widely used for lifting with mobile cranes.

The threshold for the criterion of total lifted load for mobile crane lift policies may be as low as 50,000 pounds (22,700 kilograms). This is unrealistically low for a gantry critical lift policy. Further, given the way that gantry systems function, using a criterion of total lifted load may not be meaningful at all. Four-leg gantry systems with first stage rated loads in the 400-ton to 450-ton (360-tonne to 410-tonne) range are widely available today. The use of such a system to lift a load weighing 150 to 200 tons (135 to 180 tonnes) is generally not viewed as a particularly high risk with respect to weight alone.

Much more significant than gross lifted weight is the percentage of rated load. All facets of a gantry system's rated load are generally based on structural, mechanical or hydraulic limits, unlike mobile cranes that also have stability-limited rated loads. However, before suggesting a percentage of rated load to define a critical lift, a second consideration must be examined.

A simple lift with no horizontal movement presents less demand to the system than a lift that requires travel or side shifting. Further, once horizontal movement is introduced into the operation, the extension height of the gantry legs also enters into the picture. The higher the legs are extended, the less stable the system becomes. One contractor's gantry system lift policy with which the author is familiar calls for derating the gantry legs by 10% when traveling in the first stage, by 15% when traveling in the second stage, and by 25% when traveling in the third or fourth stage.

The next step in the demands placed on the gantry system by the operation is the use of a gantry to upend or downend a load. Coordination of vertical and horizontal motions introduces additional opportunities to develop unwanted horizontal loads.

Of course, any of the advanced applications discussed in Section 2.3 must be treated as critical lifts simply because of the complexity of integrating different types of lifting equipment (e.g., strand jacks and gantries) or using the gantry system in a unique environment (e.g., on a floating barge).

One can see from this brief discussion why some consider every lift made with a gantry system to be a critical lift. However, it is more realistic to recognize that some gantry lifts are not critical lifts. The following dividing lines are suggested with respect to the weight of the load as compared to the rated load and setup of the gantry system and the presence or absence of horizontal movement.

- A lift that does not require horizontal movement and that does not exceed some specified percentage of the rated load, perhaps 75% to 80%, can be treated as a standard lift. (SC&RA 2004 suggests a threshold of 70% of the rated load for such a lift when the gantry legs will work in the upper stages.)
- A lift that requires horizontal movement, either travel or side shifting, that is performed at an extension height that is not greater than two-thirds of the maximum extension height and that does not exceed a specified percentage of the rated load, in this case perhaps only 65% to 70%, can be treated as a standard lift.

- Only lifts where the gantry track is supported on a slab on grade, on soil with suitable mats and timbers, or on a substantial structure, such as a powerhouse turbine pedestal, can be treated as standard lifts.
- Every other lift made with a hydraulic gantry system should be treated as a critical lift.

A complete critical lift policy must also add variables that address the nature of the lifted load (hazardous material, difficult to handle, unusually valuable), site conditions, and other aspects of lifting that may increase risk. Due to the current lack of formal gantry lift policies, the ideas listed above should be considered as food for thought. Individual contractors, preferably working with the manufacturers of the gantry systems they use, should expand this list to develop more refined critical lift policies.

A standard lift can be planned and performed using relatively informal means, such as just making a rough sketch to describe the equipment layout and a checklist or form to identify and evaluate that equipment. A critical lift requires a formal lift plan that addresses the issues discussed in Section 7.2.1. The specific details will vary, just as they do for a mobile crane lift plan, depending on the demands of the particular lift to be made. Regardless of the level of detail, the three required steps are: plan, communicate, execute.

7.3   EQUIPMENT SETUP GUIDELINES

Although it shouldn't need to be said (but it will be anyway), the setup of a hydraulic gantry system should, first and foremost, follow the manufacturer's requirements. The manufacturer of the hydraulic gantry system will normally specify in the operator's manual requirements for the setup and operation of the equipment. If the contractor finds that a deviation from these requirements must be made, the manufacturer should be contacted for advice.

Beyond that, following is a brief list of items to be considered in the setup of the gantry system.

1. Verify that all dimensions on the drawings that locate building columns, floor openings, and the like, are correct.

   Drawings of existing facilities, particularly older facilities, may not be available. Even if drawings are available, they may not reflect minor changes made during the original construction or renovations and additions made over time. When working in a new facility that is under construction, the contractor must verify the status of construction in the area at the time the lift will be made. Regardless of the source of the information upon which the lift plan was developed, the contractor is well advised to perform at least basic field measurements to verify the site conditions. If errors are found, the entire lift plan could be suspect.

2. Install any temporary supports required for the track. This may include compaction of soil, installation of floor shoring, laying down of timbers, mats or steel plates, installation of track stands, or leveling of the surface of a sloped concrete floor slab.

    Some of these support installation activities may require engineering support. Evaluation of soil bearing capacity or the strength of a concrete slab on grade often requires the assistance of a civil engineer to assure that the bearing surface can safely support the loads from the gantry leg. The design and selection of shoring for a floor may also require engineering assistance.

    Ultimately, the track must be level along its complete length. If the lift calls for travel over a long run, the surface upon which the track will be laid should be surveyed to locate the high point. The area can then be leveled or shims laid out based on that point.

3. Install the track beams. This is, perhaps, the most important aspect of the gantry system setup. Without a foundation that possesses the necessary strength and stiffness, the probability of a loss-of-stability accident increases dramatically.

    The track beams must be installed level in both directions within certain tolerances. In the absence of guidance from the gantry system manufacturer, SC&RA (2004) recommends that the track beam setup should meet the following tolerances:

    - Level to within 1/8 inch over a 10-foot length of track (within 3.1 mm over 3 meters).
    - Level within 1/8 inch (3.2 mm) across the width of the track from center of track beam to center of track beam.
    - The gap between the ends of the track beam top flanges should not exceed 1/16 inch (1.6 mm) if the track beams have square ends.
    - If the gantry system will travel with the load, the track runs must be parallel to one another within 1/4 inch (6.4 mm) over the full length of travel.
    - The elevations of the top surfaces of the track beams should not differ by more than 1/16 inch (1.6 mm) at the track connections.
    - The centerlines of the track should not be misaligned horizontally by more than 1/8 inch (3.2 mm) at the track connections.

    The most common gantry track assemblies have a gauge of 36 inches (914 mm) and track for some of the larger machines have a gauge of 48 inches (1,219 mm). Thus, a common (in the U.S.) 4-foot carpenter's level can easily be used to check the level of the track setup in the lateral direction. For use of such a level in the longitudinal direction, note that 1/8 inch in 10 feet equals 0.05 inch (just under 1/16 inch) in four feet.

The horizontal alignment tolerance of 1/8 inch may be excessive for some gantry/track combinations for the track beams with guide bars. The misalignment of the guide bars at the track connections should be no more than one-half of the total clearance between the guide bar and the wheels. For example, if the clear space between the gantry leg's wheels is 1.125 inches and the guide bar diameter is 1.000 inch, then the total clearance is 1.125 - 1.000 = 0.125 inch and the suggested maximum guide bar misalignment is 0.125 / 2 = 0.063 inch.

Most gantry track assemblies bolt together end to end. These connections are not necessarily designed to carry the full shear force that will occur if the joint is supported on only one side. Therefore, every track beam joint should be uniformly supported across both sides of the joint, unless the track manufacturer assures that a one-sided support is acceptable.

Installation of stops at the ends of the track beams, or preferably at the ends of planned system travel, is strongly recommended.

4. For some gantry setups, track beams may not be required. It is sometimes possible to roll gantries on steel plates laid on a concrete slab floor.

   Concrete slab floors are usually sloped for drainage, but not always noticeably so. If plate on the floor is to be used as a support surface, the same requirements for out-of-level apply here as are discussed above for track beams. As a practical matter, the use of plate under a gantry leg is typically acceptable only for the smaller legs due to the magnitudes of the wheel loads that are developed by the higher capacity systems.

5. Assure that the top surfaces of the track beams (or track plate) are clean and smooth along all areas where the legs will roll.

   This is as simple as pushing a broom along the beams or plate. The need is to remove any materials, such as small stones, that would interfere with the smooth rolling of the gantry legs' wheels or rollers.

6. Install the gantry legs on the track beams. Verify that the track is level by checking the plumb of the legs at two locations 90° apart.

   The most important point to remember when setting the gantry legs on the track is that they all must face in the same direction if the system will travel with the load. That is, "forward" must be the same for all of the legs in the system.

   The check of the plumb of the leg is a verification of the track installation work. If the track is level, the gantry leg will be plumb in both directions. Likewise, if the gantry leg is not plumb in both directions, a problem with the track setup is indicated.

7. Install the header beams on the gantry legs and, if required, the cross beams on the header beams.

The ends of the header beams should overhang the edges of the gantry leg header plates to allow easy visual verification from the ground that the beams bear across the full width of the header plates. Further, the header beams should be clamped or otherwise positively secured to the legs. If a four-beam or other such arrangement is to be used, the cross beams should be positively connected to the header beams using clamping devices, bolts, or other suitable hardware. Heavy C-clamps can be used for this purpose and there are commercial products, such as the girder clamp shown in Fig. 7.11, that are also suitable. As with the header beams, the ends of the cross beams should overhang the edges of the header beams slightly.

Note that the lift links and side shift devices, if used, must be installed on the beams before their placement on the system. The lift links and side shift devices must be secured to avoid shifting or falling during the handling of the header or cross beams.

Particular care must be taken when the header beam is relatively deep and the gantry leg header plate will allow rotation side to side (with respect to the length of the header beam). If the header plate is mounted on a rocker that pivots in one direction only with the axis of rotation perpendicular to the length of the header beam (Fig. 7.12a), then the header beam will be stable once the header plate beam clamps are installed. If the header plate is mounted on a ball-and-socket (Fig. 7.12b) or a rocker with the axis of rotation parallel to the length of the header beam, then the header beam will be free to lean to one side. Depending on the depth and weight of the header beam, the amount of rotation allowed by the header plate mounting and the design of the header plate beam clamps, this action could allow the header beam to break free of the header plate.

In such a situation, either temporary bracing must be employed or the header plate must be blocked to prevent rotation until the full beam setup is assembled. Any bracing or blocking that is used here must be removed before making the lift to assure that the header plate can move as intended by its design.

Figure 7.11 Commercial Girder Clamp Product *(Lindapter International)*

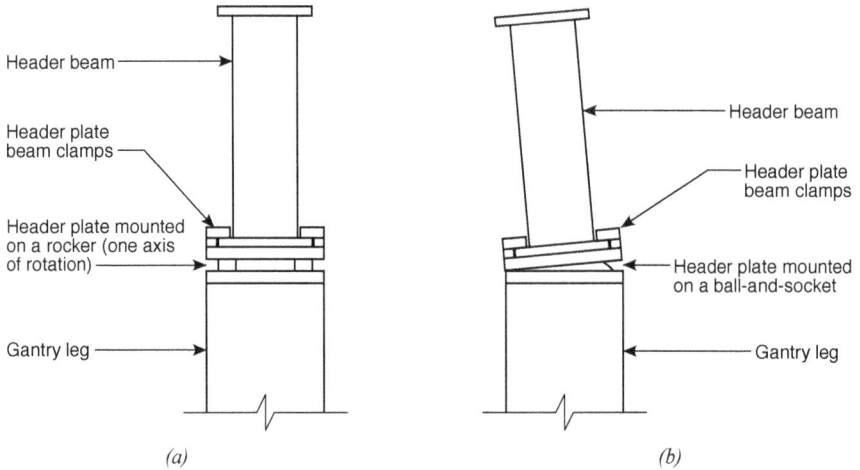

Figure 7.12 Header Beam Stability During Gantry System Assembly

8. Install all hydraulic hoses, electric control lines, external propel devices, if applicable, and other such items appropriate to the particular gantry system for the planned operations.

   There are no general guidelines that can be given here, other than the oft-repeated edict to follow the directions in the operator's manual for the system. If the controls are not hooked up correctly, the gantry system will not operate correctly. Connecting everything the right way is no more difficult or time consuming than doing it wrong. When one recognizes the magnitude of damage that can be caused if the gantry system does not respond to the operator's actions as expected, we can see the importance of getting the controls connected and functioning correctly prior to beginning a lift.

9. The use of a load-indicating device on each of the gantry legs is strongly recommended.

   Most of the newer gantry systems have control units that provide the operator with load readings for each leg. If the system being used is so equipped, the gantry operator must assure that the load-indicating devices are zeroed out or otherwise set up to provide accurate readings during the lift. If hydraulic pressure readings are to be used to monitor the load, a table equating pressure readings to loads should be mounted on the control unit. Regardless of the method or type of equipment to be used to monitor the load, the operator should know before the lift begins the expected loads or pressures at each leg.

10. Perform a system check of all hook-ups and functions in accordance with the gantry system operator's manual.

At the least, this check should cover all movements that will be required to perform the lift. It is usually not necessary to run the gantry system to the full limits of motion in each function. The primary goal here is to allow the operator to verify how the system responds to the controls and to assure that every needed function works. In some cases, however, a full dry run may be appropriate. For example, a lift that will bring some parts of the gantry system into close proximity with obstructions may benefit from a dry run to absolutely verify all clearances.

11. Install the rigging between the lift links and the padeyes, trunnions, or other lifting attachments on the load.

    This activity applies standard industrial and construction rigging practices and is generally not unique to working with hydraulic gantries.

All of the activities discussed to this point are what are needed to get ready. Now it's time to actually make the lift.

7.4 LIFTING OPERATIONS

The actual performance of the lift follows many of the normal practices of heavy rigging that are independent of the type if lifting equipment in use: Move the load slowly; continually monitor the rigging and keep it plumb; if any sign of trouble is seen, stop, determine what is wrong, and don't proceed until all concerns have been satisfactorily addressed. Additionally, the gantry legs, themselves, should be checked for plumb throughout the lift. This check should be done using a level or other measuring device. It's best not to rely on "eyeballing" the legs to verify that they are plumb. Of course, if access to the legs as required to use a level is very difficult or presents unacceptable risk, checking the plumb of the legs by sighting is better than doing nothing.

7.4.1 Performance of the Lift

The key aspects and considerations for the performance of a lift with a gantry system can be best examined by stepping through each individual operation. Of course, not every lift will involve every one of the operations discussed here, so each operation is discussed in a "stand-alone" format.

These discussions are general in content, not specific to any particular model of gantry system. All operations of a gantry system should be performed only by appropriately trained personnel. Each gantry manufacturer offers operator training courses for buyers of their systems. It is incumbent upon the contractor to assure that its field people have the necessary knowledge and training to operate the equipment correctly and safely. Remember that a gantry system is not a mobile crane and lifting with gantries is not like jacking and cribbing. Gantry system operation is a unique skill set.

The Pre-Lift Meeting. As is the case with most any lift, a pre-lift meeting should be held immediately before the lift begins. This is the last opportunity to go over the lift plan, to identify key personnel, to assure that everyone's tasks are clearly understood, to agree on the communication method to be used (B30.1 hand signals, radios, etc.), and to allow any concerns to be discussed and remedied.

Just as the lift plan for a simple lift need not be an elaborate document, the pre-lift meeting doesn't have to be a formal affair. Just gather the crew, review the lift plan, and be sure that the crew is ready. All questions must be answered and any problems or safety concerns resolved to assure that everyone is on the same page moving forward.

The Initial Lift. Making the initial lift of the load sets the stage for all that follows. This movement must be made with care and with a knowledge of what to watch for and how to recognize and correct potential problems before they become real problems.

Initially, the rigging between the load and the header beams will be slack, but installed such that the lifting attachments on the load are aligned with the lift links as required by the planned rigging arrangement. The lift is initiated by extending the gantry legs to snug up the rigging. The most important observation to make as this movement proceeds is the point at which each rigging assembly becomes taut. Remember that a 10-foot sling is rarely exactly 10 feet long. Manufacturing tolerances come into play here, potentially allowing one or two slings to snug up first.

Consider the lift of a torsionally rigid object from four lifting attachments, as shown in Fig. 7.13. This is a common occurrence in gantry work and was discussed in Chapter 4 with respect to the development of uneven loading due to cross cornering. If, for example, the sling nearest Gantry Leg 1 is a couple of inches shorter than the other slings and the two header beams are level along their lengths and with each other, this sling will tighten first. If the extension of the four legs simply proceeds, the slings nearest Gantry Legs 1 and 4 will end up carrying the lion's share of the load. This, in turn, will significantly change the distribution of the load among the four gantry legs and their supports.

Here's where the operator's skill comes into play. To simplify the example, let's assume that the slings nearest Gantry Legs 2, 3, and 4 are identical in length, the sling nearest Gantry Leg 1 is 1 inch (25 mm) shorter than the others, and that the lift links are 8 feet (2.44 meters) apart along the header beams. The operator can compensate for the sling length difference by holding Leg 1 and continuing to extend the other three legs, allowing all four slings to become taut together. The header beam spanning between Legs 3 and 4 will be level, but the header beam

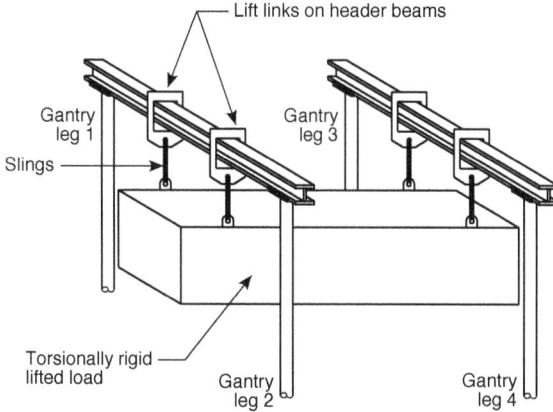

Figure 7.13 Lifting from Four Attachment Points

spanning between Legs 1 and 2 will have a small slope. Using the numbers in this example, the slope of the beam will be 1 / 96 = 1.04%, or 0.6°. This slope is small enough that it can be tolerated for many lifting operations.

If the difference in rigging lengths is great enough that the header beam slope would be intolerable, other steps must be taken, such as shimming under the lift links at the locations of the longer rigging lengths. Regardless of how the problem is addressed, the important point is that it is addressed. As we saw in Chapter 4, the distribution of the weight of the lifted load can differ considerably from the assumed balance if cross cornering is allowed to occur.

Many gantry system users (and some of the gantry manufacturers) recommend that at least some of the legs should be allowed to freewheel as the initial lift is made. For the lift arrangement illustrated in Fig. 7.13, for example, Gantry Legs 1 and 2 might be fixed, but Legs 3 and 4 would be allowed to freewheel. Some even recommend allowing all of the legs to freewheel. The purpose for doing so is to allow the legs to center themselves around the load in response to horizontal forces that may be developed due to small initial misalignments of the gantry system. Once the horizontal forces become great enough to overcome the rolling resistance, the gantry leg will shift along the track.

Additional issues to be considered are very similar to those discussed for any lift, regardless of the type of lifting equipment being used. The gantry legs should be observed for any indication of moving out of plumb and the supports (track beams, cribbing, stands, and everything else that supports the system) must be watched for any signs of distress as load is taken by the gantry legs. Of course, the lift should stop immediately if any problem is seen or even suspected so the problem can be analyzed and corrective action taken.

Another common practice calls for stopping the lift motion and holding the load for a few minutes once it is freely supported by the gantry system. This allows a quick visual check of the rigging, the gantry system, and the supporting surface before proceeding. An additional suggestion calls for lowering the load slightly to assure that the legs will retract smoothly and for testing the travel and side shift functions, if either will be used as a part of the lift, prior to raising the load more than just a few inches. Any indicated problems when making these tests are much easier to address if the load has been lifted only a few inches.

Continuing the Lift. Once the load is fully supported by the gantry system and the crew has verified that the equipment is functioning properly, the lift can proceed. What happens next depends greatly on the specifics of the particular gantry system being used.

Just as assuring that the rigging goes taut uniformly, thus assuring the expected distribution of the weight of the lifted load, the gantry legs must extend uniformly to maintain that distribution. The burden of making sure this happens rests upon the operator, particularly when using a gantry system with manual controls. Some type of monitoring should be used to check the extension lengths of the legs. This can be a simple steel tape measure fixed to the top of each leg and watched by a spotter or it can be an electronic device with a digital read-out at the control unit. Regardless of the type of hardware used, the leg extension progress should be monitored and, if necessary, stopped and corrected.

One particular point to watch is that at which the legs move from one stage to the next. If the lift arrangement works out such that all of the legs of the system start and remain at the same extension, then all of the legs will transition from one stage to the next together. This is the ideal situation, but is not always possible. If the legs do not transition together, care must be exercised by the gantry operator to be sure that the leg moving, for example, from the first stage to the second doesn't change extension rate due to the smaller bore diameter of the lift cylinder in the second stage. A constant fluid flow rate will cause the cylinder to extend more rapidly in the second stage, but a constant fluid pressure will cause the cylinder to stall. The gantry operator must be prepared to adjust the controls accordingly.

Gantries with today's more sophisticated computer-based control systems can be set up to perform all of these observations and corrections automatically. Sensors that monitor the load carried by each leg and the extension height of each leg provide input to the control unit. Software routines then process this information and adjust the flow rate and pressure as needed to maintain a smooth, constant extension of all of the legs. However, the use of such a control system does not give the operator permission to ignore the hardware. A computer system is an aid, not a replacement for a qualified gantry operator. It will always be the responsibility of the operator to assure that the gantry system is moving as desired. Continuous observation of the gantry system is still required and the operator must take over if necessary to make corrections.

Traveling with the Load. Traveling the gantry system with the suspended load should be a simple matter, provided the setup guidelines and checking of clearances discussed in Section 7.3 have been followed. There are, however, three fundamental requirements for successful travel. The first two have as their collective goal the minimization of dynamic and other horizontal forces. The third regards monitoring of clearances.

The first requirement is that of smooth operation of the propel system. Whether the system being used is a hydrostatic drive or a simple set of propel cylinders, starting and stopping motions must be very smooth, slow, and controlled in order to minimize the dynamic loads that occur during acceleration and deceleration. Once under way, the travel speed should be held constant to minimize swaying of the suspended load, again with the goal of minimizing horizontal forces. For straight travel, all of the legs on one track should be tied together with connecting members. This is usually a requirement when using propel cylinders and is advisable even when using gantry legs with internal drives (Fig. 7.14). These members assure that the legs will move as a group and prevent the out-of-plumb rigging discussed in Section 4.1.3.

The second requirement is to assure that the travel motion of the gantry legs on one track run are synchronized with the travel of the legs on the other track run(s). It is imperative that the legs on each side of the load travel in sync. We discussed in Chapter 4 the effect called racking, in which the legs on one side of the load get ahead of the legs on the other side. If allowed to become excessive, racking can generate forces to the gantry system components that can overstress the parts and even contribute to a total loss of stability. Thus, as the gantry system is traveled, the legs must be monitored to assure that one side does not get ahead of the other. The monitoring of system travel can be done quite simply with spotters tracking the gantry leg positions along the track relative to reference marks laid out as a part of the equipment setup or by laying tape measures along the tracks. As with lifting, some of the control systems currently available measure gantry leg travel and can make adjustments automatically as needed to minimize racking. And again as with lifting, these systems should not be relied upon blindly. The gantry

Figure 7.14 Connecting Member Between Legs with Internal Drives *(David Duerr, P.E.)*

operator and crew are still responsible for the safe operation of the system and must not become complacent.

Third, verification of clearances between the gantry system and the suspended load with respect to any obstructions along the travel path should have been performed as a part of the lift planning (Section 7.2) and then checked again during the equipment setup (Section 7.3). This doesn't mean that a third check isn't needed. Any time the gantry system is in motion, spotters should be watching the system and the load to assure that everything clears. Potential obstructions include structural members on either side, roof beams or an overhead crane above, and anchor bolts or other embedments that protrude upward from the floor below. Having the load or a part of the gantry system, such as an overhanging header beam, hang up on an obstruction can result in the toppling of the system. This is not something to be taken lightly.

Side Shifting the Load. The concerns during side shifting are very much like the concerns during travel. The starting and stopping motions must be smooth, slow and controlled, all of the side shift units must move together, and clearances around the equipment and the load must be monitored at all times that the suspended load is in motion. Side shifting tends to be easier to manage than travel since the overall distance of movement is typically less. Whereas a gantry system may be traveled dozens of feet, side shifting is most commonly not more than 10 or 15 feet. Thus, there is less opportunity for misalignments to develop or contact with obstructions to occur.

Even though side shifting may appear to be a more controlled operation than travel, horizontal movement of a load still brings with it a certain amount of risk, so the operation should not be taken lightly. Side shifting creates inertial forces in the lateral direction, which for most gantry systems, is the weak direction with respect to stability. This necessitates that appropriate care be exercised during all side shift activities.

Lowering the Load. Lowering a load requires as much, if not more, care than lifting the load. The retraction of the lift cylinders is a complex function due to the interaction of the fluid pressure developed by the supported load, the pressure introduced through the operation of the system, and the behavior of the counterbalance valves. Details of the performance of the hydraulic system and its individual components are covered in Chapter 3.

The primary concern during lowering is similar to one of the concerns during lifting. That is maintaining coordination of the lengths of all of the gantry legs. In the simplest arrangement, all of the legs are extended to the same length and each carries the same load. In this case, operation of the controls to bring the legs down together is not exceptionally challenging. However, many lifts differ from this ideal situation.

As is discussed in Section 3.3.3, the pressure provided to the retract port of a lift cylinder (which is the pressure to the pilot port of the counterbalance valve) to lower the load varies inversely with the magnitude of the supported load. That is, as the load that the gantry leg is supporting increases, the retract port pressure required to retract the leg decreases. This tells us that the gantry operator must carefully manipulate the control valves for each leg to assure that their motion is uniform. And as with lifting, the transition from one cylinder stage to the next must be watched to assure that the movement remains smooth and controlled, particularly when the gantry legs are not all extended to identical lengths and, thus, do not transition together.

Of course, as with the other operations discussed in this section, the monitoring of leg extension and supported load and adjustment of the motions to compensate for changes are performed automatically by some of the computer-based gantry control systems.

The most sensitive part of the lowering operation is often the last few inches. If the load being handled is a machine component and it is being set over anchor bolts, alignment must be relatively precise. The position of the load must be accurate within a couple of sixteenths of an inch (a few millimeters) to allow the anchor bolts to pass through the bolt holes in the item's base. And this alignment must be rotational, as well as lateral and longitudinal. It is not uncommon for the crew to have to drift the load slightly to obtain this alignment to allow lowering the load those last few inches. However, knowing that the load can be drifted is not permission to set up the gantry system out of tolerance or to operate the system carelessly. Drifting a load develops horizontal forces to the gantry system (see Section 4.1.3) and must not be allowed to become excessive.

Last, the final touch-down of the load should be very controlled. Whether setting a machine on its sole plates or any sort of load on a vehicle, the contact and transfer of load from gantry system to supporting surface must minimize impact so as not to damage the load or the support.

Other Operations. The operations discussed above are the four fundamental gantry operations (lift, travel, side shift, and lower). As is discussed in Chapter 2, gantry systems can be used for a variety of other, more complex operations. However, even the most complex lifts are still just compilations of lift-travel-side shift-lower. For example, upending of a load may require lifting with two gantry legs while traveling with the other two legs. Thus, we see that mastering the four basic operations is a requirement for being able to perform the more complex gantry system applications.

The increase of the demands on the gantry operator and the crew when performing a more complex lift becomes obvious in light of these discussions of the basic operations. Not only must the different types of motions (e.g., lifting with one pair of legs and traveling with the other) be monitored, but understanding how to correct when the movements get out of tolerance is more involved.

Here's where we see that there is no substitute for training and experience when performing these more complex lifts.

7.4.2   Responsibilities

Over the past decade, construction industry safety standards increasingly have been defining the responsibilities that the various parties have to assure the safe performance of a lift. This practice has obvious management value. First, once an employee's responsibilities have been defined, the training and knowledge that the employee must possess to fulfill those responsibilities becomes clear. Second, clarification of the responsibilities of the various parties can enable a more efficient allocation of personnel to a project.

The reader must recognize that the positions, responsibilities, and training discussed in this section are all related to the functional performance of the work. It is understood that craft labor agreements that exist in some areas may define the types of work that certain employees may perform in ways that conflict with the discussion here. In those situations, it is incumbent upon the employer to understand the labor agreements and the practical requirements of the job and then to make appropriate employee assignments.

The SC&RA *Recommended Practices for Telescopic Hydraulic Gantry Systems* defines three positions applicable to the planning and execution of a lift using a hydraulic gantry system. These are the Gantry Operator, the Lift Planner, and the Lift Supervisor. Section 4.7.3 of the SC&RA guide, titled "Competency Requirements," describes the basic knowledge or skills that an individual should possess to be considered competent to function in each of these positions. These descriptions imply, but do not explicitly define, responsibilities for each of these three positions.

ASME B30.1-2009 *Jacks, Industrial Rollers, Air Casters, and Hydraulic Gantries* (ASME 2009) is currently being revised to include a section on responsibilities. Although this is a work in progress, the basic structure will follow the responsibilities section added to ASME B30.5 *Mobile and Locomotive Cranes*, starting with the 2007 edition (ASME 2008). Thus, a general discussion of the B30.1 provisions can be made here.

Since B30.1 covers a variety of types of lifting and rigging equipment, reference is made to load handling equipment, abbreviated LHE, rather than "gantry," "jack," etc. As such, B30.1 defines five positions. These are the LHE Operator, the LHE Owner, the LHE User, the Load Handling Director, and the Site Supervisor. For clarity in this discussion, we will replace "LHE" with "Gantry System." The Load Handling Director is more commonly referred to as the Lift Director with respect to gantry system operations.

The Gantry System Operator is the individual who directly controls the functions of the gantry system. The responsibilities of the Gantry System Operator include reviewing the requirements of the lift before beginning any operations, understanding the operating characteristics of the gantry system, becoming familiar with the site conditions as applicable to the lift to be made, performing equipment inspections as required by the operator's manual or by Chapter 1-6 of ASME B30.1, and understanding basic load rigging procedures.

The Gantry System Owner is defined as the entity that has custodial control of the gantry system by virtue of ownership or lease. Thus, the Gantry System Owner may not be an owner, in the traditional sense of the word. Also, the Gantry System Owner will most likely be a business, rather than an individual. For example, a company that leases a gantry system and provides it to a project site is the Gantry System Owner for the purpose of the B30.1 responsibility definition. In summary, the Gantry System Owner's responsibilities include providing to the job equipment that is in sound operating condition, meets the requirements of the job as defined by the Gantry System User, is equipped with all necessary and/or requested accessories, including load charts and operator's manuals, and using only appropriately qualified individuals for the performance of inspections, maintenance and other work on the equipment for which the Gantry System Owner is responsible.

The Gantry System User is the entity that arranges for the presence of the gantry system on the job site and controls its use there. One can easily see that the Gantry System Owner and the Gantry System User may be (and often are) the same entity. The Gantry System User's responsibilities, in broad form, require that the gantry system be in proper operating condition before its first use on the job and that all personnel involved in the use, maintenance and inspection of the gantry system understand their individual responsibilities and are appropriately qualified for their assigned tasks. The Gantry System User is typically a business.

The Lift Director oversees the work performed by the rigging crew with the gantry system. As such, a fundamental responsibility of the Lift Director is to be on site during all lifting operations. Additional responsibilities include ensuring that all work area preparations have been completed, that all related safety issues, such as traffic control and personnel access, have been addressed, that one or more signalpersons have been appointed and identified to the Gantry System Operator, that the load is properly rigged, and that the Gantry System Operator is fully informed with respect to the details of the lift plan.

The Gantry System Operator, the Lift Planner, and the Lift Director commonly work for the Gantry System User. The Gantry System Operator and the Lift Director are typically employees; the Lift Planner may be an employee or a subcontractor.

The Site Supervisor exercises supervisory control over the job site and over the work that is being performed with the gantry system. The Site Supervisor may be a representative of the owner of the facility at which the work is being performed or may be an employee of a general contractor or other entity for which the lift is being made. The responsibilities of the Site Supervisor tend to cover "big picture" issues, such as work area preparation, access, scheduling and coordination with other activities on the job site, and ensuring that the Gantry System User's personnel are all properly qualified for the jobs they are performing.

The Infrastructure Health & Safety Association's *Construction Multi-Trades Health and Safety Manual* (IHSA 2007) chapter on lifting with hydraulic gantry systems includes a brief section on responsibilities. The entities addressed in this publication are the Owner/Constructor, the Contractor, the Manufacturer/Supplier, and the Contractor/Tradespersons.

The Owner/Constructor is either the owner of the facility or a general contractor that is managing the overall construction project at the facility. A representative of this entity is the equivalent of the B30.1 Site Supervisor and is assigned similar responsibilities.

The Contractor is the lifting/rigging contractor that is performing the work with the gantry system. This entity is equivalent to the B30.1 Gantry System User and, again, is assigned similar responsibilities.

The Manufacturer/Supplier is the entity that is providing the gantry system to the project. This entity is equivalent to the B30.1 Gantry System Owner and is assigned similar responsibilities. An interesting difference, though, is the reference to the gantry system manufacturer, a party not addressed in B30.1. From a practical point of view, however, routine use of a gantry system at a work site doesn't generally involve the gantry system manufacturer. Those responsibilities assigned to the Manufacturer/Supplier, such as providing the equipment in proper working order and providing the necessary documentation, are quite reasonably expected of the Gantry System Owner.

Last, the Contractor/Tradespersons are the actual employees who will perform the work. The responsibilities of the B30.1 Gantry System Operator and Lift Director overlap with those listed for these individuals.

In summary, although the descriptive titles are different, the responsibilities listed in the IHSA guide are very similar to those given in ASME B30.1.

BS 7121-13:2009 *Code of Practice for Safe Use of Cranes – Part 13: Hydraulic Gantry Lifting Systems*, published by the British Standards Institution (BSI 2009) follows form and also includes sections on personnel responsibilities and qualifications. This standard covers the HGLS (hydraulic gantry lifting system) Lift Supervisor (Lift Director), HGLS Operator (Gantry System Operator), HGLS Slingers/Signallers, and Maintenance Personnel. Discussion of responsibilities for these last two job functions is unique to this standard.

HGLS Slingers/Signallers are the riggers and signalpersons who will work with the gantry system. The need to provide competent employees in these positions is addressed in the B30.1 responsibilities, but there are no specific responsibilities or qualifications assigned to these individuals. The responsibilities assigned in BS 7121-12 include installing the rigging, monitoring the lifting operation, and signalling the Gantry System Operator.

Maintenance Personnel are exactly what the title implies: Those who are responsible for maintaining the gantry system in accordance with the manufacturer's maintenance manual. Under B30.1, the Gantry System Owner is responsible for ensuring that maintenance is performed as required by appropriately qualified personnel, effectively providing the same coverage of responsibilities.

The qualifications requirements for these four groups of individuals as given in BS 7121-13 are quite extensive and cover not only knowledge and training, but physical fitness. (While having the physical ability to perform the required tasks each position demands is generally logical, how these requirements might conflict with provisions of the U.S. Americans With Disabilities Act is beyond the scope of this discussion.)

The full text of the provisions discussed in this section are significantly more detailed than has been presented, so those responsible for the planning and execution of gantry lifting work must obtain copies of the applicable documents to assure that their operations comply with the appropriate requirements.

7.4.3   Special Applications

The discussions in this section are fairly general in nature and oriented toward the more common applications of hydraulic gantry systems. The demands placed on both the equipment and field personnel when using a gantry system in one of the more specialized applications discussed in Chapter 2 may be much greater than the norm.

Special applications include using a gantry system in conjunction with another type of lifting equipment, such as strand jacks, using a gantry system as a tailing rig, or using a gantry system on a floating barge. The primary driver that differentiates these operations from the more conventional uses is that these applications can subject the system to forces that were not considered in the design of the gantry legs or accounted for in the methods normally used to design the system appurtenances, such as header beams and track assemblies.

These differences create a demand for a commensurate increase in the qualifications those involved must possess and the responsibilities they must bear. The Lift Planners must not only understand the principles that apply to gantry systems, but they must also understand the principles that apply to the associated equipment. For example, a Lift Planner who is designing a lift to be made on a barge, in addition to having an expertise in gantry lifting, must also have a working knowledge of naval architecture in order to evaluate the movement of the barge to changing load positions, to calculate the forces on the barge from water flow, and to evaluate the hydrostatic stability of the floating barge.

Likewise, the Gantry Operator, even if responsible for operating only the gantry system, should still have a general understanding of the other lifting equipment in order to know what to expect as the lift proceeds. When using a gantry system as a tailing rig, for example, the Gantry Operator should understand the functions of the top lifting rig, be it a mobile crane, strand jacks on a tower system or some other type of lifting equipment, in order to better control the gantry system in coordinated movement.

To put this in a more straightforward manner, when planning and performing a complex or specialized lift with a gantry system, the lifting contractor has to bring its A team. If outside people or new hires have to be brought in to staff the project with personnel with the right skill sets, so be it. Complexity brings risk and risk must be countered with skill and knowledge. This must be recognized by the lifting contractor's management and acted upon accordingly.

7.5   RISK MANAGEMENT

Hydraulic gantry systems are used to lift and otherwise manipulate heavy objects. Work of this nature always brings with it a certain amount of risk. It is the responsibility of the contractor to manage that risk in such a way that the work is performed safely and efficiently. Protecting the well being of people and property is an obvious moral imperative. Promoting safety and efficiency on the job site also brings financial rewards. As the saying goes, if you think safety is expensive, take a look at the cost of an accident.

The opportunity for something to go wrong lurks around every corner. This section breaks those opportunities down into six groups in which deficiencies may exist that can elevate the level of risk. Within each group is a list of issues or events that can impair the performance of the lift.

7.5.1   Lift Planning and Engineering

Lift planning and engineering together form the first line of defense against the occurrence of an accident and provide the foundation upon which efficient field operations are performed. The extent of planning that is required for a particular lift is a function of the complexity of that lift and the consequences of a failure. Following are discussions of four particular areas that can cause problems in the engineering and planning phase of a project.

Inadequate knowledge or training of those individuals charged with performing the lift planning creates the risk that errors will occur in the planning and engineering. With respect to overall lift planning, a certain amount of hands-on practical experience is necessary to understand how a lift is performed in the field. As previously noted, lifting with gantries is not lifting with a crane and it is not jacking and cribbing. The employer, therefore, must provide opportunities for inexperienced employees to get out onto the job site and observe firsthand how gantry systems operate and how they are used to perform a lift. With respect to engineering, one must remember that lift planning, especially with specialized equipment like hydraulic gantry systems, is not typically a part of the traditional university curriculum in civil or mechanical engineering. This means that knowledge of lift engineering is developed primarily through on-the-job training. This experience can be enhanced through education at industry-specific conferences and workshops. The SC&RA Crane & Rigging Workshop, which is held annually, is a good example of such an educational event.

Lift planning and engineering should always be performed in accordance with the appropriate industry standards and guidelines. In the United States, ASME B30.1-2009 *Jacks, Industrial Rollers, Air Casters, and Hydraulic Gantries* (ASME 2009) and the SC&RA *Recommended Practices for Telescopic Hydraulic Gantry Systems* (SC&RA 2004) must be followed. Comparable standards and guides in other countries, such as BSI (2009) in the United Kingdom, must be followed when working in those locales.

Failure to abide by accepted industry practices can allow problems to creep into the work and, in the event of an accident, can open the door to liability if the contractor is shown to have ignored or deviated from generally accepted practice. The OSHA General Duty Clause (29 USC 654 Sec. 5) states that an employer must provide a workplace that is free from recognized hazards. Even though these industry standards and guides are not regulations that are directly enforceable, their existence effectively defines a variety of recognized hazards. Thus, an accident that can be shown to have occurred as a result of not abiding by the guidance contained in these industry publications may be considered to be a violation of the General Duty clause. (This last comment applies to work sites in the United States. Similar regulations in other countries must, of course, be followed as appropriate.)

When the gantry track system bears on soil, both the loading applied to the soil and the bearing capacity of the soil (and the strength of subsurface pipes and structures) must be determined during the planning phase in order to allow the design of adequate supports for the track. This work may require the assistance of a geotechnical engineer to investigate subsurface conditions and determine the allowable ground bearing pressure for the lift site. This is not something to be left for the field crew to sort out. Excessive settlement of the track beams, particularly when unequal from one support point to the next, can significantly reduce the stability of the gantry system. When coupled with other conditions of the lift, such as development of lateral forces during side shifting, this reduction of stability can result in the toppling of the system.

When the gantry track system is supported on a structure, such as a building floor, the track support loads and the ability of that structure to safety carry those loads must be evaluated as a part of the lift planning. As with soil support, excessive deflection of floor beams or other structural supports will diminish the stability of the gantry system. Performance of this activity requires detailed knowledge of the structure and an appropriate level of structural engineering expertise. Again, this work cannot be left for the field crew to figure out.

7.5.2   Work Site Preparation

This set of activities encompasses the work that is performed from the time the crew arrives on the job site to the beginning of the setup of the gantry system, itself. In this context, the "gantry system" consists of the track assemblies, the gantry legs, and everything supported by the legs (e.g., header beams, lift links, rigging, etc.).

If the work site is not prepared in accordance with the requirements of the lift plan, proper setup and operation of the gantry system may be impossible. Improper preparation may be driven by schedule pressure, inadequate crew staffing, or unqualified supervision. Specific areas in which deficiencies may creep into this part of the project are discussed below, with suggestions on how to reduce the risk associated with each.

First, and perhaps most important, is making a check of the work site conditions. Any deviations from the expected conditions may invalidate the planning that has been performed up front. Therefore, the area must be checked for access, clearances, ground conditions (if the track is to be set up on soil), the configuration and condition of the building structure (if working on a floor), and any other characteristics that may affect the ability to execute the lift as planned. Even if not a safety issue, changes in site conditions can have significant commercial ramifications for the contractor and the owner, so any differences found in site conditions should be carefully documented.

One condition that can change in a way that significantly affects the bearing capacity of soil is the presence of ground water. The ground water elevation will change over time in response to rain, ice melt, and other natural activities. Underground water can also appear unexpectedly as a result of a leaking pipe. If the gantry system will be set up on bare ground or on a slab on grade, a part of the mobilization check of the work site should include looking for signs of subsurface water.

The surface on which the gantry track will be set must be prepared in accordance with the lift plan. This may include cleaning the floor or slab on grade of debris that will impair the leveling of the track, grading and compacting soil, locating and marking underground utilities to assure that supports do not bear directly over these items, and any other preparatory work required by the lift plan. This work is followed by placing timbers, mats, shims, and other load carrying materials that will support the track beams. As has been discussed at length elsewhere, accurate setup of the track is absolutely essential for safe operation of the gantry system. The quality of this work must be high and should be verified by the Lift Director before continuing with laying the track beams. Compensating for errors in this work may be impossible down the road.

We have discussed the importance of equal supports where the track beams must span over a pit or between structural supports. Both beams of each track assembly must be supported identically to assure that they will behave identically during the lift. While the Lift Planner and Lift Director should understand this requirement, others in the crew may not have this knowledge. The Lift Director must check the work as it proceeds to assure that all special track support requirements are met. The risk here is created by a lack of experience on the part of the craft labor. The compensation for this deficiency is experience and supervisory action on the part of the Lift Director.

As the gantry system equipment arrives on site, the hardware should be inspected for missing parts and damage. One part that often does not receive the respect that it should is the operator's manual. Every gantry system should have an operator's manual so that any questions regarding the setup and operation of the system can be answered accurately and quickly. Related to the need for an operator's manual is the need to have a load chart. Whether the load chart is a printed sheet mounted on the control unit or an electronic chart built into the control system, the rated load information must be readily available to the Gantry System Operator. Ideally, the load chart should show the rated load per leg, not just of the gantry system as a whole with the load equally shared among the legs.

7.5.3   Gantry System Setup

The assembly of the gantry system, from the placement of the track beams to the positioning of the lift links, must meet the requirements of the gantry manufacturer and the demands of the job. Errors here can simply make the execution of the lift a little more difficult or they can create insurmountable obstacles.

The tolerances to be met in the placement of the track beams are discussed on page 219. Errors in this work are typically created by a simple lack of care in laying out the track assemblies and then in checking track level and alignment as the work proceeds.

The problems created by poor track installation can be significant. Track that is out of level in the longitudinal direction forces the gantry system to move uphill or downhill as the system travels. Even if travel isn't required, gantry legs that are supported on an out of level surface are less stable. This reduction in stability is even more pronounced if the track is out of level in the lateral direction. If the two runs of track along which the gantry system travels are not parallel, the legs can be forced together or apart as the system moves. Since the tops of the legs are typically fixed relative to one another by the header beams, this action can induce bending in the legs and create a lateral force that can reduce stability. Last, misalignment of the joints from one track beam to the next will impair smooth travel of the legs as the wheels or rollers are forced across the joints.

Fine debris on the rolling surfaces of the track beams, such as sand or grit, will hinder the smooth travel of the gantry legs. Larger obstructions, such as a large chunk of weld spatter, can completely stop leg travel. Thus, before the gantry legs are placed on the tracks, the top surfaces of the track beams should be swept off and then checked for debris. A smooth track surface is a necessity.

Many gantry systems must have the legs connected to the control and power unit by hydraulic hoses or electrical cables. Careless routing of the hoses or cables can result in damage which, in turn, can abruptly impair the operation of the gantry system. Hoses or cables of inadequate length can be pulled taut during travel or side shifting; if this action isn't observed in time, they can even be pulled from their connections.

Orientation of the gantry legs as they are placed on the track must be consistent so the travel function works correctly. That is, "forward" must be the same for all of the legs in the system. The same is true for side shifting devices, when used. Although errors in this part of the equipment setup will be caught when making a dry run (if a dry run is performed), it is always best to avoid such errors up front.

Placement of the header beams and cross beams on the gantry legs can be a rigging project in itself. The lift links are normally installed on the beams before the beams are lifted into place, so the links must be secured to prevent shifting during the handling of the beams. Tall header beams, such as fabricated beams that are much deeper than they are wide, can fall over if handled roughly. Header beams should be secured to the header plates as soon as they are set and bracing should be used if needed to maintain stability of the beams (Fig. 7.12). Likewise, cross beams should be secured to the header beams as soon as they are set.

When working indoors with a gantry system that uses an internal combustion engine for power, exhaust gases present a safety hazard to personnel. Setup of the system must include the installation of ducting to vent the engine's exhaust to the outside of the building.

7.5.4   Performance of the Lift

As has been said before, here's where the rubber meets the road. Even when a comprehensive, well written lift plan is in place, all of the equipment is in top operating condition, and the gantry system setup conforms with all of the project requirements, care must still be taken in the operation of the system in the performance of the lift.

Most of the accidents that have occurred with gantry systems ultimately resulted in the toppling of the system. This tells us that reducing or eliminating actions that develop horizontal forces is of paramount importance when performing the lift. Actions that can produce horizontal forces and, therefore, must be minimized, include the following:

- Rigging between the load and the lift links is not vertical.
- Rigging goes out of plumb significantly while upending or downending a load (complete elimination of out of plumb rigging in this type of operation is almost impossible; some out of plumb of the rigging must be expected and accounted for in the lift plan).
- Pendulum action of the load occurs due to poorly controlled starting and stopping of either travel or side shifting movements.
- Misalignment of the gantry system develops during travel (racking).
- High wind load occurs due to performance of the lift during unacceptably high wind speeds.
- Excessive drifting of the load is required to align the load for setting (this is a product of poor track setup).

The extension and retraction of the gantry legs of a manually operated system may require special attention on the part of the Gantry System Operator if all of the legs do not carry the same load. Unequal load distribution is not at all uncommon and typically results from picking a load that is offset in one or both horizontal directions relative to the center of the gantry system (e.g., Fig. 7.6). If the system is not properly controlled, either manually or by means of a computer-based control system, the legs may not extend or retract as required, resulting in the header beam going out of level. The remedy to this is the chain of lift planning, informing the Gantry System Operator of what to expect in terms of loads to each leg as the lift proceeds, and attention on the part of the Operator to control the system appropriately. Having spotters observe the legs while raising or lowering the load is very desirable.

The last set of risks to be considered here regard the interaction between the gantry system and the surroundings. Clearances between all parts of the gantry system, including the suspended load, and structures and equipment in the vicinity must be monitored at all times that the gantry system is in motion. No part of the system can be allowed to hang up on or otherwise contact any obstruction. Such an interference can damage the items that come into contact or, at worst, cause enough of a reaction that the gantry system overturns. Clearances between the gantry system and electrical lines must also be monitored. When working outdoors, the risk is overhead power lines, just as when working with mobile cranes. When working indoors, electrical conduits, overhead crane buss bars, and other high voltage electrical gear can be damaged by contact with, for example, the end of an overhanging header beam. Such contact can also result in the energizing of the gantry system with potentially deadly results.

7.5.5   Dismantling of the Gantry System

The dismantling of a gantry system is not necessarily the reverse of its assembly. Depending on the nature of the lift that has just been completed, disassembly may have to proceed while working over and around the newly set load. Consequently, methods that were suitable for assembly, such as using a forklift to set the header beams on the retracted legs, may not be an option for disassembly. This point shows the importance of having a plan for the dismantling of the gantry system after completion of the lift. And last, the cautions discussed on page 221 about placing and securing deep header beams on the header plates apply equally to the disassembly of the system.

7.5.6   Equipment Inspection and Maintenance

Just as lift planning and engineering form the first line of defense against the occurrence of an accident, inspection and maintenance of the gantry system form the first line of defense against an equipment malfunction. All of the industry standards and guides that we have discussed here require regular inspections of the equipment to uncover developing problems and performance of maintenance to keep the equipment in acceptable operating condition.

Listing here the various inspections and maintenance activities that are required or recommended is not particularly practical, due to the many differences from one type of gantry system to the next. So we fall back on the often repeated directive to follow the requirements spelled out by the manufacturer and described in the system operator's manual. The requirements of Section 1-6.8 of ASME B30.1 should also be followed.

One issue that is often overlooked is the effect of the weather, particularly the temperature, on the operation of a gantry system. As discussed in Chapter 3, the viscosity of hydraulic fluid changes with the temperature. The lower the temperature, the higher the viscosity and, therefore, the greater the resistance to flow. There are different grades of hydraulic fluid available, each designed for a different operating temperature range. A gantry system should be operated with a fluid that is suitable for the work environment.

7.5.7   Quantifying Risk

Many organizations have developed methods by which the risk level of a lift to be made with mobile cranes can be quantified. These methods typically require the lift planner to answer a series of questions on a flow chart (e.g., "Is the lifted load over 50,000 pounds?"). By working through the flow chart questions, the planner is led to conclude that the lift is a standard lift, a critical lift, or some other intermediate classification of lift. Other methods require the planner to assign a numerical value to various aspects of the lift, such as how difficult the load is to handle (unusually delicate, very flexible, etc.) or if there are any unusual constrictions at the site. These values are then used to fill out cells in a matrix or added up to create an overall score for the lift.

Some risk evaluation methods address only safety-related issues, such as lifted weight as a percentage of the crane's rated load. Other methods also address non-safety issues, such as the value of the lifted item or the delay to the project if the item is damaged and must be replaced. The effect of an accident on the public's opinion of the facility owner is considered in some risk evaluation methods.

Regardless of the specifics applied, most of the risk evaluation methods have one characteristic in common. At some point, they generally require the planner to exercise experience-based judgment. Given the complexities of some lifting operations, this is not only reasonable, it is desirable. This also shows that the evaluation of the level of risk present in a proposed lifting operation, whether using mobile cranes, hydraulic gantries or some other type of lifting equipment, is not something that can be blindly calculated, but must be assessed through the eyes of experience.

At present, there are no widely used risk evaluation procedures applicable to hydraulic gantry systems. Developing such a procedure is not practical at present due to a lack of standardization in the gantry industry. As a point of comparison, mobile cranes are designed, manufactured, and tested in accordance with a body of engineering standards that have been in use for many decades. Mobile crane operators (in the United States) are increasingly certified in accordance with uniform industry standards and mobile cranes are regularly inspected following requirements detailed in OSHA regulations and industry standards. Consequently, the construction and maintenance industries know exactly what performance to expect from mobile cranes and the people who operate them.

The design and manufacture of hydraulic gantry systems, on the other hand, are not subject to comparable engineering standards and there are no operator certification programs for gantry operators beyond the training provided by the individual gantry system manufacturers. Thus, gantry system and operator performance cannot necessarily be expected to be as uniform as for mobile cranes. Lacking this uniformity in the equipment, the development of a "one size fits all" risk evaluation procedure for lifting with hydraulic gantries is impractical.

The author is familiar with gantry lift evaluation procedures developed by individual contractors, based on the contractors' specific equipment and experience. This is a practical and responsible approach to quantifying and managing the risk associated with lifting with hydraulic gantries. The development and use of such in-house procedures should be considered by contractors who use these systems. As this equipment becomes increasingly standardized, development of a broader risk evaluation method will become practical.

7.6 REFERENCES

American Society of Civil Engineers (ASCE) (2002), SEI/ASCE 37-02, *Design Loads on Structures During Construction*, Reston, VA.

American Society of Civil Engineers (ASCE) (2010), SEI/ASCE 7-10, *Minimum Design Loads for Buildings and Other Structures*, Reston, VA.

American Society of Mechanical Engineers (ASME) (2008), ASME B30.5-2007 *Mobile and Locomotive Cranes*, New York, NY.

American Society of Mechanical Engineers (ASME) (2009), ASME B30.1-2009 *Jacks, Industrial Rollers, Air Casters, and Hydraulic Gantries*, New York, NY.

APA – The Engineered Wood Association (formerly the American Plywood Association) (APA) (1998), *Plywood Design Specification*, Tacoma, WA.

British Standards Institution (BSI) (2009), BS 7121-13:2009 *Code of Practice for Safe Use of Cranes – Part 13: Hydraulic Gantry Lifting Systems*, London, U.K.

Infrastructure Health & Safety Association (IHSA) (2007), M033 *Construction Multi-Trades Health and Safety Manual*, Mississauga, Ontario.

Specialized Carriers & Rigging Association (SC&RA) (2004), *Recommended Practices for Telescopic Hydraulic Gantry Systems*, Centreville, VA.

Appendix 1
Glossary of Specialized Terms

Annulus   The space between the inside surface of a cylinder's barrel or sleeve and the outside surface of the next inner sleeve or the rod. Also called annular space.

Annular Area   The cross-sectional area of the annulus, perpendicular to the axis of the cylinder. Fluid pressure acts on the annular area to retract a double acting cylinder.

Annular Space   See Annulus.

Bare Cylinder Gantry Leg   A gantry leg that is configured with one or more exposed lift cylinders and without a lift boom. The lift cylinders provide structural resistance to all forces, both vertical and horizontal, acting on the gantry leg.

Barrel   The outermost section of a hydraulic cylinder.

Base Section   The outermost, stationary structural section of a telescopic boom.

Base Weldment   The structural weldment that forms the main base housing of the gantry leg.

Boom   See Lift Boom.

Check Valve   A directional control valve that allows free fluid flow in one direction, but completely blocks flow in the opposite direction.

Communication Holes   Holes in a cylinder's rod or sleeves that permit the flow of fluid between annuli.

Connecting Member   A load carrying element that is used to connect together two gantry legs positioned on the same track. A connecting member may be a structural member, a hydraulic cylinder, or other item capable of transmitting the propel force from one leg to the other. Although a chain or similar item can be used as a connecting member, an element that can transmit forces in both tension and compression is usually preferred.

Control and Power Unit   A remote station from which the gantry system is operated. On an all-hydraulic gantry system, the control and power unit contains an electric motor or internal combustion engine, an oil tank, and a hydraulic pump that provides oil flow to control levers on the control and power unit. This flow is transmitted to the gantry legs through pairs of hoses. On an electric-over-hydraulic system, the control unit contains electrical controls that are used by the operator to send signals to an external power unit or the gantry legs which contain the motor, hydraulic pump, and oil tank. On this type of system, only electric cables connect the gantry legs to the control unit. On both types of systems, radio remote controls may also be used. Also called a control module. See Control Unit, Power Unit.

Control Module   See Control and Power Unit.

Control Unit   A remote station that contains controls only that is used to operate a gantry system. See Control and Power Unit, Power Unit.

Counterbalance Valve   A pressure control valve installed in a hydraulic circuit to control flow in a specified direction while allowing free flow in the opposite direction; for example, to control the retraction of a cylinder when the cylinder is supporting a load. Also referred to as a holding valve or a lock valve.

Cracking Pressure   The pressure at which a pressure control valve, such as a relief valve or a counterbalance valve, begins to open and pass fluid.

Cross Beam   A structural section, usually a wide flange rolled shape or a fabricated box section, that spans between header beams and to which the lifted load may be attached, typically using lift links. See Header Beam, Lift Link.

Cross Cornering   The condition in which a suspended load intended to be supported by four points is actually supported in such a manner that two diagonally opposite support points carry a disproportionately large share of the total load. The upper limit of cross cornering occurs when the total load is supported on two diagonally opposite points.

D-Ring   See Lift Link.

Design Factor   The ratio of the limit state stress(es) or strength of a structural or mechanical element to the permissible internal stress(es) or force(s) created by the external load(s) that acts upon the element.

Directional Control Valve   A flow control valve that allows the gantry system operator to direct hydraulic fluid flow to perform a particular function (for example, to select between extending or retracting a cylinder). See Pressure Compensated Valve.

Downend   The rotation of a load from a vertical orientation to a horizontal orientation using lifting equipment. Also called laying over. See Tripping, Upend.

Drift   Horizontal displacement of the top of a telescopic lift boom that results from clearances between boom sections at the overlaps. Drift is a product of geometry (length of boom sections, length of overlaps, and clearances) and is not to be confused with structural deflection of the boom.

Drifting   Pushing or pulling a suspended load to change its horizontal position.

Dynamic Load   A load that is the result of acceleration or deceleration, usually related to a lifting operation.

Extend Port   Line attachment port on a cylinder into which pressurized fluid is introduced to extend the cylinder. See Retract Port.

External Power Unit   A gantry system power unit that is a stand-alone component. See Control and Power Unit, Internal Power Unit, Power Unit.

Freewheel   To roll freely, with neither drive power nor braking restraint provided by mechanical devices.

Gantry Leg   The assembly of the base weldment, most commonly mounted on wheels or rollers, one or more lift cylinders or a lift boom, and the mechanical, electrical and hydraulic components required to perform the leg's functions. Gantry legs with wheels or rollers may be freewheeling or self-propelled. The lift cylinders or lift booms of the gantry legs (two or four gantry legs are usually operated as a set) are extended or retracted to lift or lower the load. Wheel- or roller-mounted gantry legs can move longitudinally to transport loads horizontally. Also called a jacking unit, lift housing, lifting housing, or lifting unit.

Gantry System Operator   The individual who directly controls the functions of the gantry system.

Gantry System Owner   The entity that has custodial control of the gantry system by virtue of ownership or lease.

Gantry System User   The entity that arranges for the presence of the gantry system on the job site and controls its use there.

Gauge   Center-to-center spacing of the beams of a track assembly; center-to-center spacing of the wheels or rollers of a gantry leg.

Gland Cap   A fitting at the open end of a cylinder's barrel or sleeve that contains one or more rings and seals to contain the fluid pressure within the cylinder. See Stop Ring.

Header Beam   A structural section, most commonly a wide flange rolled shape or a fabricated box section, that spans between two gantry legs and is supported on the gantry legs' header plates. Rigging may be attached to the header beam, typically using lift links, and to the load to facilitate lifting or lowering the load. Alternately, the header beams may support one or more cross beams. See Cross Beam; Header Plate; Lift Link.

Header Plate   An adapter plate that is attached to the top of the lift cylinder(s) or lift boom by means of an articulating connection. The header beam is supported on the header plate.

Height Indicator   A device installed on the gantry leg that measures the stroke and/or overall height of the lift boom or lift cylinder.

Holding Valve   See Counterbalance Valve.

In-Service Wind   The maximum wind speed that may be encountered while a gantry system is in use; typically the basis for the determination of wind loads acting on the gantry system and lifted load during system operation. See Out-of-Service Wind.

Inertial Load   See Dynamic Load.

Internal Power Unit   A gantry system power unit that is a part of the gantry leg. See Control and Power Unit, External Power Unit, Power Unit.

Jacking Unit   See Gantry Leg.

Lateral Direction   The horizontal direction perpendicular to the length of the gantry system track. See Longitudinal Direction, Side Shifting.

Lay Over   See Downend, Tripping.

Lift Boom   An assembly of two or more coaxial structural tubes that telescopes as the load is lifted or lowered and that provides structural strength and stiffness to resist horizontal forces that may occur during lifting. These tubes are most commonly round, rectangular, or octagonal, and may be fabricated or rolled shapes. The telescoping of the boom is controlled by one or more internal or external hydraulic cylinders (see Lift Cylinder). The lift boom may also provide vertical support to the lifted load by means of devices such as cams, pins, or wedges that mechanically lock the sections of the boom to one another.

Lift Cylinder   A hydraulic cylinder, either single stage or multiple stage (telescopic), that is mounted on the gantry leg base weldment and serves to raise and lower the lifted load. Lift cylinders are most commonly double acting (pressure extend, pressure retract).

Lift Director   The individual who oversees the work performed by the rigging crew with the gantry system.

Lift Housing   See Gantry Leg.

Lift Link   A device that is used to provide a point of attachment between the header or cross beam(s) and the rigging that is attached to the load. The most basic lift link is a flat plate that is cut out to fit around the header or cross beam and drilled at the bottom to allow attachment of a shackle or other pin-connected fitting. More complex lift links that provide a means to side-shift the load are also in use. The basic plate lift link is commonly referred to as a D-ring. See Side Shift.

Lifting Attachment   A load supporting device that is attached, usually by bolting or welding, to the item being lifted and that is used for connection of rigging for the purpose of lifting. Lifting attachments include lifting lugs, padeyes, trunnions, and similar appurtenances.

Lifting Capacity   See Rated Load.

Lifting Housing   See Gantry Leg.

Lifting Unit   See Gantry Leg.

Load   With respect to the performance of a lifting operation, the item being lifted (e.g., the lifted load); with respect to lift planning and engineering, an applied force (e.g., wind load).

Load Point   The position on the header or cross beam where a supported load is applied. The location of the load point is most commonly described by measuring from the centerline of the gantry leg to the load point.

Load Sensors/Gauges   Devices used to measure the total weight being carried by each gantry leg. Load sensors and/or gauges may be installed on the control unit or on each gantry leg.

Loading Point   See Load Point.

Lock Valve   See Counterbalance Valve.

Longitudinal Direction   The horizontal direction parallel to the length of the gantry system track. See Lateral Direction, Propel.

Manual Boom Section   A section of a telescopic lift boom (most commonly the uppermost section) that is extended by the leg's cylinder and then mechanically locked into position to allow retraction of the cylinder and re-engagement of the retracted cylinder for an additional stroke. See Power Boom Section.

Offset Load   A load in which the center of gravity is not at the geometric center of its supports.

Out-of-Service Wind   The maximum wind speed that may be encountered while a gantry system is fully assembled, but not rigged to a load or in operation; the basis for the determination of wind loads acting on the gantry system when installed but not in operation, if required. See In-Service Wind.

Overturning Moment   The effect created by a horizontal force acting at some height above the track (usually at the top of the gantry leg) that acts to overturn (topple) the gantry leg.

P-Δ Deflection   The increase in the horizontal displacement of the top of a gantry leg due to the P-Δ moment. See P-Δ Moment.

P-Δ Moment   The bending moment in a gantry leg created by the applied axial load (P) and the horizontal displacement (Δ) of the top of the leg relative to a vertical line through the base of the leg. (Note that this behavior is nonlinear since a change in the P-Δ moment results in a change in the P-Δ deflection, which is a part of the overall deflection Δ.)

Piston   A fitting at the inside end of a cylinder's rod or sleeves that contains one or more rings and seals to contain the fluid pressure within the cylinder and that provides a solid surface against which fluid pressure acts to extend the cylinder.

Plunger   See Rod.

Power Boom Section   A section of a lift boom that is extended and retracted by means of one or more hydraulic cylinders during the lifting operation. See Manual Boom Section.

Power Unit   An assembly that contains an electric motor or internal combustion engine, an oil tank, and a hydraulic pump that provides oil flow to the gantry legs.

The power unit may be a stand-alone unit (external power unit) or built into the gantry legs (internal power unit). See Control and Power Unit.

Pressure Compensated Valve   A flow control valve that provides a constant rate of fluid flow regardless of load pressure upstream from the valve.

Propel   Longitudinal movement of the gantry system along the track (used as either a noun or a verb). Also called travel.

Propel System   The hydraulic and mechanical components within the gantry system that provide the drive (propel) force for longitudinal travel. There are three basic concepts for propel systems in common use.

1. Built-in (or Integral) Drive: The gantry leg contains a motor that drives some or all of the wheels or rollers of the leg through a gear train or chain drive.
2. External Cylinder: Used on freewheeling gantry legs, a cylinder is pinned close to the bottom of the leg's base at one end and to the track at the other. The cylinder is extended and retracted to move the gantry legs along the track.
3. External Drive Wheel: One or more motor-driven drive wheels are mounted to the base of the gantry leg. Tractive effort is developed by pre-loading the wheel against the track to create adequate friction.

Propelling Gantry Leg   The longitudinal movement of a gantry leg. Also, the mechanism utilized to power this longitudinal movement. (This term is defined in the SC&RA *Recommended Practices for Telescopic Hydraulic Gantry Systems*. However, the terms Propel and Propel System, defined above, are more commonly used in practice.)

Punching Shear   The loading of a flat structural element, such as a concrete floor slab or a steel plate, in the through-thickness direction. A punching shear failure occurs due to excessive shear stress created by a local concentrated load applied to one surface of the element.

Racking   A longitudinal misalignment of a gantry system that occurs during travel when the legs on one track move ahead of the legs on the other track. (The term Skewing is used when referring to this type of behavior in overhead bridge, gantry, or portal cranes.)

Rated Capacity   See Rated Load.

Rated Load   The maximum load for which a lifting device may be used under specified working conditions. Also called lifting capacity or rated capacity.

Relief Valve   A pressure control valve that limits the fluid pressure to a specified maximum value at a point in a hydraulic system.

Resisting Moment   See Stabilizing Moment.

Retract Port   Line attachment port on a cylinder into which pressurized fluid is introduced to retract the cylinder. See Extend Port.

Rigging   The hardware used to attach the lifted load to the lift links. Commonly used rigging items include shackles, master links, and wire rope, synthetic or chain slings.

Righting Moment   See Stabilizing Moment.

Rod   The innermost telescoping element of a hydraulic cylinder. Also called the plunger.

Safety Factor   See Design Factor.

Shuttle Valve   A control valve that connects the fluid flow from one of two inlet ports to the outlet port. The selected inlet port is that port at which the greater fluid pressure is acting.

Side Shift   The lateral movement of a suspended load along the header or cross beams, usually accomplished by means of lift links equipped with wheels, rollers, or sliding surfaces and some form of motive power. Lift links used for side shifting may be self-powered or be moved by means of external hydraulic cylinders or winching devices.

Site Supervisor   The individual who exercises supervisory control over the job site and over the work that is being performed with the gantry system.

Skewing   See Racking.

Sleeve   The inner moving section(s) of a telescopic hydraulic cylinder, other than the rod. See Rod.

Stability   The ability of a gantry leg or gantry system to resist overturning.

Stabilizing Moment   A moment created by the vertical loads acting through the horizontal distances from those loads to the tipping fulcrum. The stabilizing moment acts to resist overturning. Also called resisting moment or righting moment.

Stop Ring   A band welded to the outside of a cylinder's rod or sleeve to limit the extension of the rod or sleeve. The stop ring functions by coming to bear against the inside surface of the gland cap. See Gland Cap.

Stroke   The linear extension or retraction movement of a hydraulic cylinder or a gantry leg.

Telescopic Boom Gantry Leg   A gantry leg that is constructed with a lift boom. The lift boom provides structural resistance to all horizontal forces acting on the system and the lift cylinder(s) provide only vertical support to the lifted load.

Telescopic Cylinder   A hydraulic cylinder composed of a barrel, rod, and at least one sleeve such that extension occurs in two or more stages.

Telescopic Hydraulic Gantry System   Two or more gantry legs, a control and power unit, hoses, cables and other control accessories, and one or more header beams that span between the legs.

Tilt Up or Tilt Down Operation   Term used by OSHA in 29 CFR 1926 Subpart CC to refer to upending or downending of a load. See Downend, Lay Over, Tripping, Upend.

Tipping Fulcrum   The horizontal line about which a gantry leg will rotate should it overturn. The total weight of the gantry leg and the supported load will be exerted on this line at the moment of overturning.

Toppling   Overturning of a gantry leg.

Track   An assembly consisting of two parallel beams, usually tied together at two or more regular intervals. The beams are most commonly wide flange shapes or fabricated sections, and are typically fitted with one or more elements (wheel guide bars) that guide the gantry legs' wheels or rollers. The center-to-center beam spacing is equal to the gantry leg gauge. Track is typically designed in short sections that can be bolted or pinned together end to end. The track beams can be designed to carry the gantry leg wheel loads over significant clear spans or may be designed to be fully supported by the underlying surface, functioning only to provide a smooth, guided surface for the gantry legs' wheels or rollers. When used with the external cylinder propel device, the track assembly must also be fitted with the propel cylinder anchor.

Travel   Longitudinal movement of the gantry system along the track. See Propel.

Tripping   The rotating of a load from a vertical orientation to a horizontal orientation or from a horizontal orientation to a vertical orientation using lifting equipment; or, only rotating of a load from a vertical orientation to a horizontal orientation. The use of this term should be avoided since its meaning is not consistent in all geographic regions and industries. See Downend, Upend.

Upend   The rotation of a load from a horizontal orientation to a vertical orientation using lifting equipment. See Downend, Tripping.

Wind Load   The force produced on a body by the movement of the wind, assumed to act horizontally.

Appendix 2
USCU / SI Conversion Factors

Following are conversion relationships for the most commonly encountered US customary units (USCU) and International System of Units (SI) quantities applicable to the use of hydraulic gantry systems. The standard abbreviations for the SI quantities are shown in parentheses. These factors are as defined in IEEE/ASTM SI 10-2010 *American National Standard for Metric Practice*. A multiplier in bold type indicates that the conversion factor is exact and, therefore, all subsequent digits are zero.

To convert from	To	Multiply by
Length		
inches	millimeters (mm)	**25.4**
feet	meters (m)	**0.304 8**
yards	meters (m)	**0.914 4**
miles (U.S. statute)	kilometers (km)	1.609 347
Area		
square inches	square millimeters (mm^2)	**645.16**
square feet	square meters (m^2)	**0.092 903 04**
Volume		
cubic inches	cubic millimeters (mm^3)	**1.638 706 4 E04**
cubic feet	cubic meters (m^3)	0.028 316 85
quarts (U.S. liquid)	liters (l)	0.946 352 9
gallons (U.S.) (231 in^3)	liters (l)	3.785 412
Mass		
pounds	kilograms (kg)	**0.453 592 37**
short tons (2,000 pounds)	metric tonnes[1] (t) (1,000 kg)	0.907 184 7
Force		
pounds	newtons (N)	4.448 222
short tons (2,000 pounds)	kilonewtons (kN)	8.896 444
kilograms	newtons (N)[2]	**9.806 65**

To convert from	To	Multiply by
Bending Moment		
pound-inch	newton-meter (N-m)	0.112 985
pound-foot	newton-meter (N-m)	1.355 818
Pressure or Stress		
pounds per square inch	kilopascal (kPa)[3]	6.894 757
pounds per square foot	pascal (Pa)	47.880 26
Power		
horsepower (550 ft-lb/s)	kilowatts (kW)	0.745 699 9
foot-pounds/second	watts (W)	1.355 818
Temperature		
degrees Fahrenheit (°F)	degrees Celsius (°C)	t °C = (t °F - 32) / 1.8
Unit Weight		
pounds per foot	newtons per meter(N/m)	14.593 90
pounds per foot	kilograms per meter (kg/m)	1.488 164

[1] The metric ton, or tonne, is not a formal SI unit, but is in common usage in heavy lifting and rigging work.

[2] This conversion is based on the acceleration due to gravity, equal to $9.80665 \, m/s^2$.

[3] 1 bar = 100 kilopascals.

INDEX

www.ingramcontent.com/pod-product-compliance
Lightning Source LLC
Chambersburg PA
CBHW050456190326

41458CB00005B/1305